Tropical Forests: A Call for Action

Part I
The Plan

Report of an International Task Force convened by the World Resources Institute, The World Bank, and the United Nations Development Programme

World Resources Institute

October 1985

Each World Resources Institute study represents a significant and timely treatment of a subject of public concern. WRI takes responsibility for choosing the study topics and guaranteeing its authors and researchers freedom of inquiry. It also solicits and responds to the guidance of advisory panels and expert reviewers. Unless otherwise stated, however, all the interpretations and findings set forth in WRI publications are those of the authors.

Peter Smith
Production Coordinator

Magazine Group
Design

Library of Congress Catalog Card Number 85-51864
ISBN 0-915825-10-4

Contents

World Resources Institute gratefully acknowledges the financial support provided for this project by The World Bank, the United Nations Development Programme, the U.S. Agency for International Development, the Canadian International Development Agency, the Netherlands Government, and the W. Alton Jones Foundation.

Foreword

Tropical forests are one of the earth's most valuable natural resources. Throughout history, they have been essential sources of food, fuel, shelter, medicines, and many other products. They sustain people and their environments by protecting soil and water resources and providing habitat for an estimated 50% of the world's plant and animal species. It is likely that tropical forests also influence regional and global climate.

Because tropical forests benefit people in so many ways, the alarming rate of forest destruction should be a matter of grave concern. Every year more than 11 million hectares—an area larger than Austria—is lost.

The lives of more than one billion people in the developing countries, primarily the rural and urban poor, are disrupted by periodic flooding, fuelwood scarcity, soil and water degradation, and reduced agricultural productivity—all caused in whole or in part by deforestation. Scientists estimate that 40% of the biologically-rich tropical moist forests have been cleared or degraded already. In many developing countries they will all but disappear in two or three decades if present trends continue.

Despite this grim prognosis for tropical forests, the basis for hope is strong. Deforestation can be arrested and, ultimately, reversed. Decades of experience have demonstrated many successful solutions to deforestation and related land misuse. However, these efforts have been isolated and far too small to address the problem. There must be greater political awareness of deforestation's negative impacts on human welfare and the environment, and the political will to mobilize all necessary human and financial resources to do something about it.

This positive conviction spurred the World Resources Institute (WRI), in cooperation with the World Bank, the United Nations Development Programme, and bilateral aid agencies, to launch this major initiative in tropical forest conservation and development. *Tropical Forests: A Call for Action* is the report of a WRI Task Force of nine world leaders in agriculture, forestry, and conservation.

This report contributes to the continuing efforts of the United Nations Food and Agriculture Organization (FAO) to raise political awareness of the action needed to combat tropical deforestation. FAO, which declared 1985 the "International Year of the Forest," is preparing a Tropical Forests Action Programme under the direction of its Committee on Forest Development in the Tropics. The FAO Action Programme, which provided the framework for the WRI effort, identifies five priority areas for action:

Fuelwood

Forestry's role in land use

Forest industrial development

Conservation of tropical forest ecosystems

Institution strengthening: research, training, and extension.

The WRI Task Force focused on translating known solutions and strategies into a five-year program of accelerated action (1987–91) that would lay the groundwork for longer term investment. Examples of successful projects are presented that illustrate the range of solutions available. Based on these success stories, and the lessons learned from past failures, priority areas for investment and action are proposed. Major policy issues and constraints that need to be addressed to carry out this program are reviewed.

The examples of successful projects and the lessons they provide are presented as case studies in Part II of the report. Part III presents five-year investment profiles for 56 countries.

The result, we believe, is a path-breaking report that recognizes the great challenges posed by tropical deforestation while offering a positive and practical strategy for its amelioration. *Tropical Forests: A Call for Action* is not just about trees, but about people and their prospects for a better life. We fervently hope that this report will help stimulate financial and policy commitments from developing and industrial country leaders, development assistance agencies, and the private sector for a greatly expanded and coordinated global effort to combat deforestation.

Helpful comments and suggestions were received from over a hundred individuals and organizations. We deeply appreciate these efforts and know that the final report benefited greatly from them.

James Gustave Speth
President
World Resources Institute

In this report, the term "tropical forests" refers to forests in the humid and semiarid/arid areas of developing countries. Thus, the term includes forest formations ranging from moist (or closed) tropical forests to dry (or open) woodlands. In a few instances, developing countries with temperate forests are also included.

A Call for Action

Summary statement of the Task Force

Onc-fourth of humanity lives in poverty, characterized by poor health, malnutrition, chronic deprivation, and shortened lives. This suffering and wasted potential is one of the great tragedies of modern times. A contributing cause of this desperate situation is the widespread loss of forests taking place throughout the developing world. Tragically, it is the rural poor themselves who are the primary agents of destruction as they clear forests for agricultural land, fuelwood, and other necessities. Lacking other means to meet their daily survival needs, rural people are forced to steadily erode the capacity of the natural environment to support them.

The relationship between poverty and deforestation is clear, but it is not inevitable. Many solutions to this cycle of increasing misery are known and demonstrated. Given appropriate government policies and institutional support, much can be done now through well designed forestry and agricultural investment programs. The evidence is clear that forest conservation and development projects can earn high enough rates of economic return to be self-sustaining. The challenge is to put these solutions to work for the millions of rural poor seeking a better future.

To meet this challenge requires political leadership. The failure of national governments and the international community to respond adequately to the deforestation crisis has led to extremely high costs in developing countries. Much of the environmental damage, decline in agricultural productivity, and human suffering that developing countries are facing today could have been reduced or avoided by greater political commitment to forest conservation and development.

To quote from a recent report of the World Commission on Environment and Development:

> Long range programmes that would have helped to tackle the underlying problems have received comparatively little support. The anti-desertification programme adopted by the UN in 1977, for example, was largely ignored by donor and recipient governments alike. That programme, it is interesting to note, was estimated to cost US$4.5 billion per annum to the year 2000 for the entire globe. A breakdown of this figure reveals that the estimated cost for Ethiopia was US$50 million per year to the year 2000. Neither the political will nor the money could be found to implement this programme, however. Yet, eight years later, faced with a human drama beyond precedent, the world community has had to find an estimated US$ 400 million for crisis-response measures to date for Ethiopia alone, and this figure will undoubtedly exceed US$500 million before the next harvest. It will go well beyond that if the harvest fails again. The arithmetic of prevention is almost always persuasive; somehow we have to invent a politics of prevention that can match the politics of crises.

This Task Force initiative seeks to advance the argument for increased action against deforestation from the narrow confines of the forestry community to the wider arena of public policy. The primary audience includes political leaders and decisionmakers in national governments and development assistance agencies who can influence policies and allocation of resources to promote the conservation and sustainable development of tropical forests.

This initiative is the first time that major development assistance agencies have helped develop a priority action program to address deforestation issues on a broad front. Investment needs for a five-year action program are proposed for 56 countries. These are preliminary estimates for consideration by developing country governments and the development assistance community.

Forestry activities alone cannot reverse current trends in deforestation. The causes of the crisis are rooted in agriculture, energy, and other sectors as well as in forestry—and so must be the solutions. Also, a broad effort involving the public and private sectors—from development assistance agencies to government ministries to local community groups—is needed.

The Task Force estimates the level of public and private investment needed to make an impact on tropical deforestation over the next 5 years to be US$8 billion. About US$5.3 billion (two-thirds of the total) would be needed for the 56 seriously affected countries reviewed in this report.

At least 30% of the proposed investment would be agriculture-related. The main goal would be to provide farmers and landless people living in or adjacent to threatened tropical forests, in overpopulated uplands, and semiarid areas with alternatives to destruction of forests and woodlands.

Half of the total US$8 billion, or US$800 million each year for 5 years, would need to be mobilized by the development assistance agencies and international lending institutions, with the remainder coming from the private sector and national governments.

Investment of US$800 million a year in forestry and related agricultural development would double the present levels of external aid in these areas.

These sums are not small—except in relation to the returns. The initiatives proposed here will alleviate hunger and deprivation, arrest dangerous assaults on the planet's environmental support system, and provide the basis for sustainable economic growth. By any system of accounting that can encompass true costs and benefits, the investment required is nothing more than a small down payment on a far brighter future.

The Task Force

The high costs of deforestation

Deforestation in developing countries

Developed and developing countries differ sharply in the condition of their forests and the status of forest conservation and management.[1] The forest area of many developed countries has stabilized and, in some cases, has increased during this century.[2]

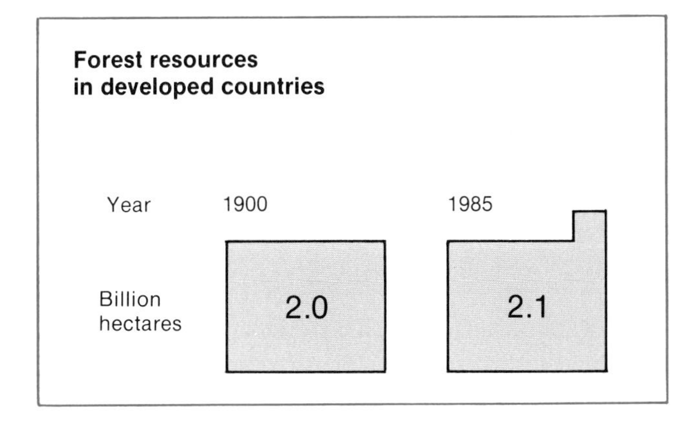

This situation has not always been so. Europe and North America have both suffered severe deforestation in the past, resulting in environmental degradation and human hardship. Fortunately, political leaders became aware of the negative impacts of deforestation and took corrective action. Principles of forest management were developed and put into practice. Laws and other institutional mechanisms were established to promote sustainable use of forest resources.

In addition, overall economic growth and expansion of job opportunities outside of agriculture led to migration of rural populations to cities and towns; this sharply reduced the numbers who depend directly on agriculture for a living and eased the pressure on forest land. The intensification and increased efficiency of agricultural production further reduced the pressure on forests. In many areas, abandoned agricultural lands reverted naturally to forest.

[1]In this report, the term "tropical forests" refers to forests in the humid and semiarid/arid areas of developing countries. Thus, the term includes forest formations ranging from moist (or closed) tropical forests to dry (or open) woodlands. In a few instances, developing countries with temperate forests are also included.

[2]Although the problem of acid rain and other air pollutants is posing a serious threat to forests in some temperate areas.

Having achieved a reasonably stable forest resource base, forest policy planners in the developed world have turned their attention to maximizing forest productivity. Well established forestry institutions now exist in most developed countries. With enough wood production to meet most of its needs for timber, plywood, and paper, the developed world maintains many forests solely for their recreational, protective, and aesthetic values.

In contrast to the developed world, forests in the developing countries have declined by nearly half in this century. Each year more than 11 million hectares of tropical forests are being cleared for other uses—7.5 million hectares of closed forests and 3.8 million hectares of open forests. In most countries, the deforestation rate is rising. If this continues, at least 225 million hectares of tropical forests will be cleared by the year 2000.

Forty percent of the closed tropical forests have been cleared, logged, or degraded. Most of the remaining 800 million hectares are in the Amazon and Congo basins, where they survive largely because of their vastness and relative inaccessibility.

However, even in countries such as Brazil where the national deforestation rate is relatively low, large areas of closed forests have been cleared in several parts of the country. Open forests—distinguished from closed forests by their discontinuous canopy and substantial grass layer—have also suffered extensive degradation.

Deforestation is a complex problem. The spread of agriculture, including crop and livestock production, is the single greatest factor in forest destruction. The rural poor are often unjustly held responsible. They are often the instruments of forest destruction, caught in a chain of events that forces them into destructive patterns of land use to meet their basic needs for food and fuel. The real causes of deforestation are poverty, skewed land distribution (due to historical patterns of land settlement and commercial agriculture development), and low agricultural productivity.

These factors, combined with rapid population growth, have led to increasingly severe pressure on forest lands

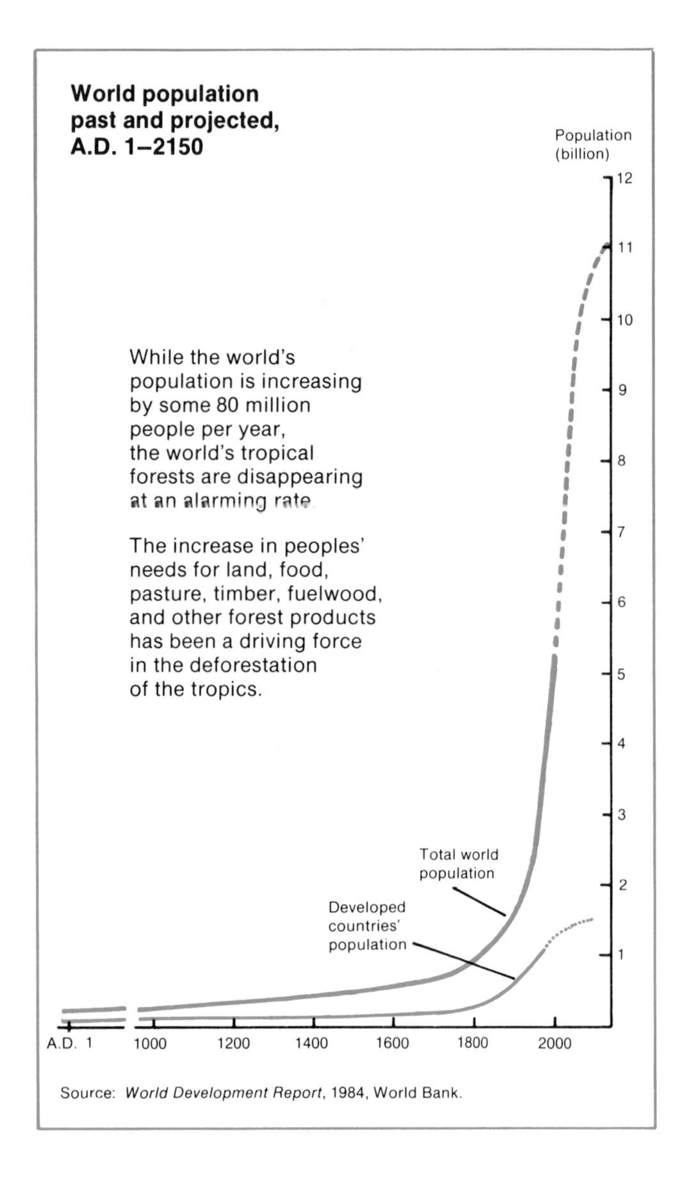

Source: *World Development Report*, 1984, World Bank.

in developing countries. As productive land becomes scarce, small farmers have been pushed into fragile upland forest areas and marginal lowlands that cannot support large numbers of people practicing subsistence agriculture. The loss of forests and rising population pressure have forced farmers to shorten fallow periods, degrading the productive capacity of the land and setting in motion a downward spiral of forest destruction. This situation prevails now in many developing countries, and it can change only if rural populations are given alternatives to this ecologically destructive way of living.

Government policies have contributed to depletion and destruction of tropical forests. Many developing countries have forestry policies (such as direct subsidies and lenient forest concession terms) that foster "mining" and unsustainable use of forest resources. Similarly, agriculture, land settlement, and other nonforestry policies often lead to encroachment on forests far beyond what is economically justified or environmentally sound.

Developed countries must share the blame for the plight of forests in developing countries. Developed country demand for tropical timber has been rising steadily. For many developing countries desperate to earn foreign exchange to ease their international debt problems, forests represent a ready source of income. A related problem is the generally low price paid for tropical timber. When prices are too low to fully reflect the growing and replacement costs for forests, there is little incentive to manage the resource for the long term. This results in the "cut-and-run" pattern of commercial forest exploitation practiced in many developing countries.

Given the history of forest exploitation in developed countries, some ask why developing countries should not follow the same path. On the surface, this is a fair question, but major differences between the two situations preclude such a strategy. The most serious difference is human numbers. The pressures of rapidly increasing populations in most developing countries completely alters the context of forest land use.

This does not mean that forests in developing countries should be left untouched. Forests are valuable resources that can provide myriad benefits to people and support economic growth. The real issue is how these resources are put to use. Currently, forest exploitation in most areas is unsustainable. In effect, a renewable resource is being treated as a nonrenewable resource.

Wood scarcity, declining food production, and desertification

Dependence on forests and trees

Almost 70% of the people in developing countries, most of whom live in rural areas, depend mainly on wood to meet their household energy needs. Low incomes restrict their ability to buy any type of fuel, so these families use wood, crop residues, dry dung, twigs, grass, or whatever source of energy can be freely gathered.

The importance of forests and trees extends far beyond their value as a source of fuelwood. Forests and trees provide wood for building poles, furniture, roof timbers, fencing, household implements, and many other uses.

Nonwood resources are also vitally important. Trees are an essential source of fodder for livestock. They also provide fruits and nuts, honey, gums, oils, resins, medicines, tannins, fibers, and other materials. There is growing recognition of the importance of small-scale forest-based enterprises as a source of nonfarm employment and income.

Forests and trees also contribute to agricultural production. In the tropics, trees do not necessarily compete with food crops, and they are often an integral part of farming systems. Trees can play a vital role in sustaining crop yields by—

- Helping maintain the soil and water base for agricultural production, particularly in upland watersheds, by reducing erosion and moderating stream flows
- Restoring soil fertility in shifting agriculture
- Increasing farm crop yields by 20-30% in arid and semiarid areas by slowing wind and increasing soil moisture
- Increasing soil nitrogen content through use of leguminous nitrogen-fixing tree species
- Providing a significant proportion of livestock feed requirements, especially in upland and semiarid regions.

As deforestation progresses, it reduces the quality of life of millions of people in developing countries. For the poorest, living close to the land, their survival is threatened by the loss of the vegetation upon which they depend. As trees disappear, so do their source of household energy and many other goods. Worse, a chain of events is set in motion that leads to declining food production, land degradation, and, in extreme cases, desertification.

The fuelwood crisis

More than 80% of the wood harvested in developing countries is used as fuelwood, compared with less than 20% in developed countries.

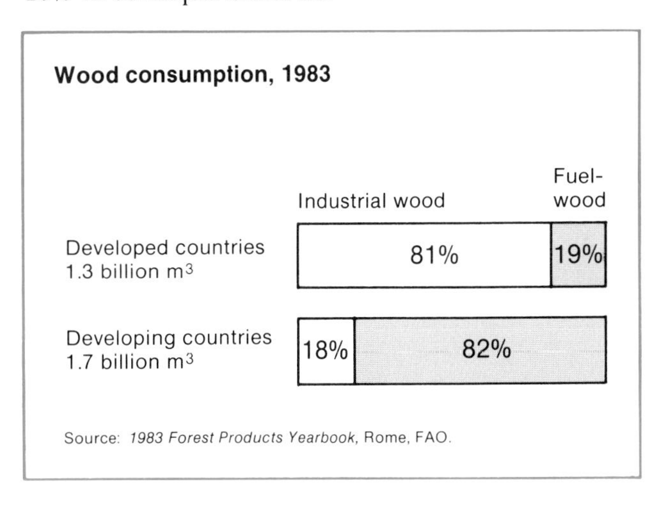

Source: *1983 Forest Products Yearbook*, Rome, FAO.

Developing countries rely on forests to meet half of their total needs for energy. In Africa, 76% of the total energy consumed is supplied by fuelwood.

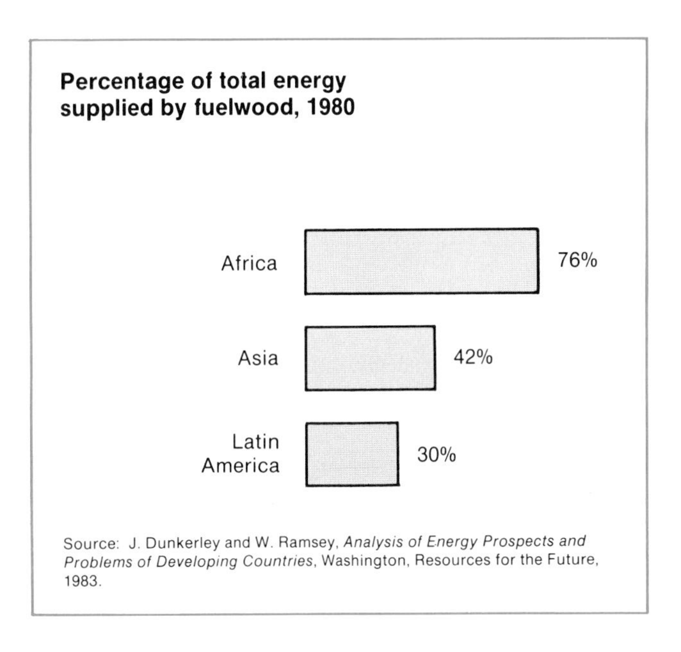

Source: J. Dunkerley and W. Ramsey, *Analysis of Energy Prospects and Problems of Developing Countries*, Washington, Resources for the Future, 1983.

In rural areas, gathering and transporting fuelwood increasingly dominates the daily lives of millions of people—100 to 300 workdays each year must be devoted to supplying a household. Women and children shoulder

As deforestation spreads, the burden of collecting fuelwood worsens for this Ethiopian woman.

most of the burden for finding and carrying home the wood needed to cook the day's meals. In Nepal, groups of villagers must leave at sunrise in order to return by sunset with a backbreaking load of wood that will last only 3 to 4 days. The increasing time needed to collect fuelwood is disrupting family stability and shortens the time that can be devoted to weeding and tending crops, preparing food, and other domestic activities.

In urban areas, most households must buy fuelwood or charcoal. Prices have risen so sharply in recent years that in many areas the wood used for cooking costs more than the food being cooked. Between 20 and 40% of the cash income of the average urban household must be set aside to buy wood or charcoal. In some countries, malnutrition is due not to lack of food but to the lack of fuelwood for cooking. Families are forced to eat less nutritious quick-cooking foods or even uncooked meals to an extent that impairs their health. Urban demand for fuelwood and charcoal is expanding the economic distance for clearing and hauling wood, leading to ever-widening circles of devastation around cities and towns.

A recent FAO analysis indicates that 1.5 billion people (70% of the 2 billion who rely on fuelwood to meet a major part of their household energy needs) are cutting wood faster than it is growing back. Some 125 million people in 23 countries cannot find enough wood to meet their needs, even by overcutting the forests.

Without major policy changes to ensure better fuelwood conservation and increased supplies, by the year 2000 some 2.4 billion people (more than half the people in the developing countries) will face fuelwood shortages and will be caught in a destructive cycle of deforesta-

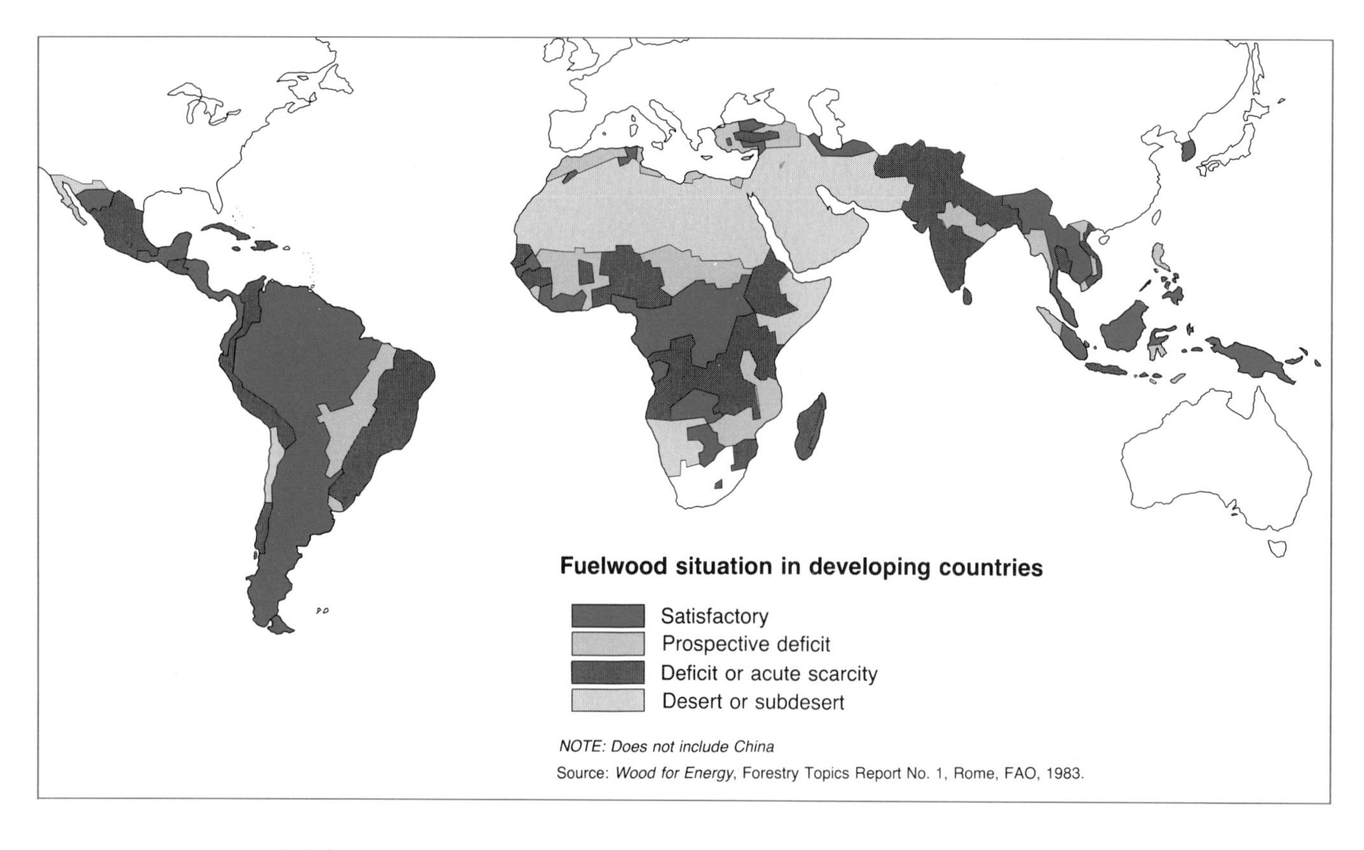

tion, fuelwood scarcity, poverty, and malnutrition.

Declining food production and desertification

Deforestation is having serious impacts on food production. As fuelwood supplies are depleted, families turn to whatever substitutes are available, primarily crop residues and animal dung. Their use as fuel robs farm fields of badly needed organic matter and nutrients. The failure to renew soil fertility leads inevitably to declines in crop yields.

The annual burning of an estimated 400 million tons of dung to cook meals in areas where fuelwood is scarce decreases food grain harvests by more than 14 million tons. This loss in the food supply is nearly double the amount of food aid annually provided to developing countries.

The removal of tree cover can further reduce agricultural productivity by loss of the benefits trees provide for farms. Widespread loss of vegetation reduces the effectiveness of rainfall by decreasing the amount of water that percolates into the ground. Water runoff increases, erosion accelerates, the water table is lowered, and springs and wells dry up.

In its most extreme form, deforestation leads to "desertification"—a process of decline in the biological productivity of arid, semiarid, and subhumid lands (or drylands). The result is desert. Drylands are particularly sensitive to human abuse because of the fragility of the soil and low and erratic rainfall. Traditional production systems are breaking down in these areas under the combined pressures of population growth and poverty. As drylands are stripped of woody vegetation through agricultural clearing, overcutting for fuelwood, overgrazing, and bush fires, land degradation worsens and the spread of deserts accelerates.

Desertification is undermining the food-producing capacity of drylands in Africa, Asia, and Latin America. A 1984 assessment by the United Nations Environment Programme shows that desertification is spreading, affecting more and more land and people. Some 1.3 billion hectares are at least moderately desertified in these three regions, and more than 300 million people live in areas at least moderately or severely desertified. The most critical areas in terms of the number of people affected are rainfed croplands, where desertification is accelerating in all three regions. Forestry and agriculture have a vital role in preventing the spread of deserts and in recovering some of the marginal areas already abandoned.

In India, animal dung made into cakes is being burned in ever greater amounts as fuelwood supplies dwindle, robbing croplands of a major source of nutrients. This is reducing crop yields.

Degraded upland watersheds

Upland and lowland populations within a watershed depend closely on one another. On upland hills, violent tropical rainstorms require close protection of the soil by vegetation. For this purpose, forest cover is best, but contour-planted tree crops can be effective. Annually cultivated crops expose bare soil and need full protection by terracing.

Skillful land use that maintains the environmental stability of upland source areas of streams not only benefits upland inhabitants but can also protect downstream hydropower reservoirs and irrigation systems from silt and debris. Erosion and sedimentation of rivers are kept to natural levels. This minimizes the raising of river beds and spreading of floods, which bring damage and misery to those living below. In return, the hill populations depend on the wealth generated by the larger communities in the valley bottoms and plains for the provision of roads and other services.

Despite their critical importance, an estimated 160 million hectares of upland watersheds in the tropical developing countries have been seriously degraded. Increasing population pressures and destructive land use have resulted in the loss of fuelwood and fodder supplies, greater flood damage, intensification of drought, sedimentation of dams and reservoirs, and loss of crops and livestock. This has caused unnecessary poverty in the hills and unnecessary damage to the lowlands. More than one billion people in the developing world are hurt by this process.

The Himalayan Range

This region contains the world's most severe watershed problems. On the lowland plains of Pakistan, India, and Bangladesh, over 400 million people are "hostage" to the land-use practices of 46 million hill dwellers. In India alone, the costs of the increasing flood damage and destruction of reservoirs and irrigation systems by sediment from misused slopes have averaged US$1 billion a year since 1978. India now spends US$250 million a year in compensation and damage-prevention measures. There is vast potential for generating hydroelectric power in the Himalayan region that could harness wealth from the watersheds, but investment in reservoirs is unsound because of the threat of sedimentation.

Himalayas — The world's most critical watersheds

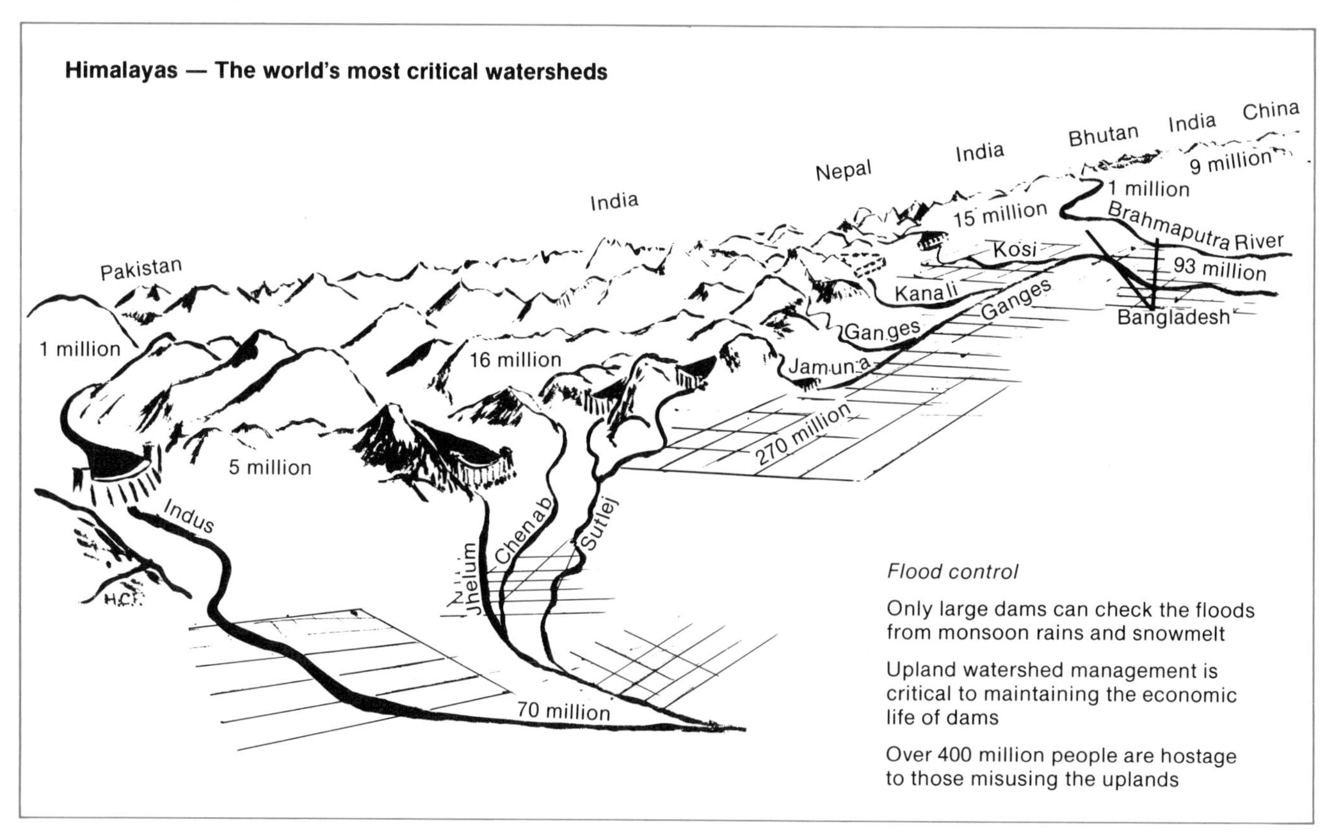

Flood control

Only large dams can check the floods from monsoon rains and snowmelt

Upland watershed management is critical to maintaining the economic life of dams

Over 400 million people are hostage to those misusing the uplands

The Andean Range

The eastern plains below the steep foothills of the Andes are typically infertile and sparsely populated, but the foothills are heavily settled and overgrazed. In the foothills, watershed problems caused by land misuse are serious—from Venezuela (where the problem is recognized as acute), through Colombia (where rehabilitation has begun), to Argentina (where clay eroded from the overgrazed watershed of the Bermejo River is carried 1200 kilometers by the Paraná River to the sea at Buenos Aires). The 80 million tons of sediment lost each year from the Bermejo watershed requires costly dredging to maintain access to the port.

The Central American Highlands

Upland watersheds in Central America are undergoing extensive deforestation, mainly for cattle raising and agriculture. Land misuse after the loss of forest cover, and a general failure to use proper soil conservation techniques, are leading to widespread soil erosion and land degradation in almost all the watersheds in the region.

The problems are most serious on the steeply sloped Pacific watersheds where most of the population lives and most of the region's food is produced. In many areas, soil erosion has become so severe that the productive potential of the land is being destroyed. Increasing rates of sedimentation threaten present and planned hydropower development throughout the region and are damaging coastal mangrove forests and fisheries.

The Ethiopian Highlands

The Central Highlands Plateau in Ethiopia supports 22 million farmers (70% of the population) and contains 59% of the country's cultivable land. Exhaustive farming practices, overgrazing, and fuelwood collection have severely eroded the plateau and destroyed most forest. Loss of soil fertility is widespread and the use of fertilizer is so limited that food production has not kept pace with population growth. Drought has precipitated a major famine.

Each year, some 1.6 billion tons of topsoil are lost from denuded lands in Ethiopia.

China's Loess Plateau

Enclosed by a bight of the Yellow River in its middle reaches, the Loess Plateau has been subject to soil erosion on a scale that is unique to China. Erosion has carved the plateau into steep rounded hills and gorges, and roads and bridges have been swept away by torrents and landslides. South of the Great Wall in this region, erosion caused by overcultivation and neglect of the poverty-stricken rural areas has reduced 100,000 hectares of fertile land to uninhabited wasteland. China already has nine people per hectare of cropland, and it needs this land.

Declining industrial wood supplies

Industrial forest products such as sawnwood, plywood, and paper are important throughout the world. They are a source of essential building materials and of the paper needed for schoolbooks, newspapers, and packing. Sustained development of the Third World implies a steady increase in demand for forest products as literacy increases and as the needs for housing, furniture, paper, and other wood-based industries grow.

Developing countries possess nearly half the world's closed forests, but they produce only 21% of its industrial timber.

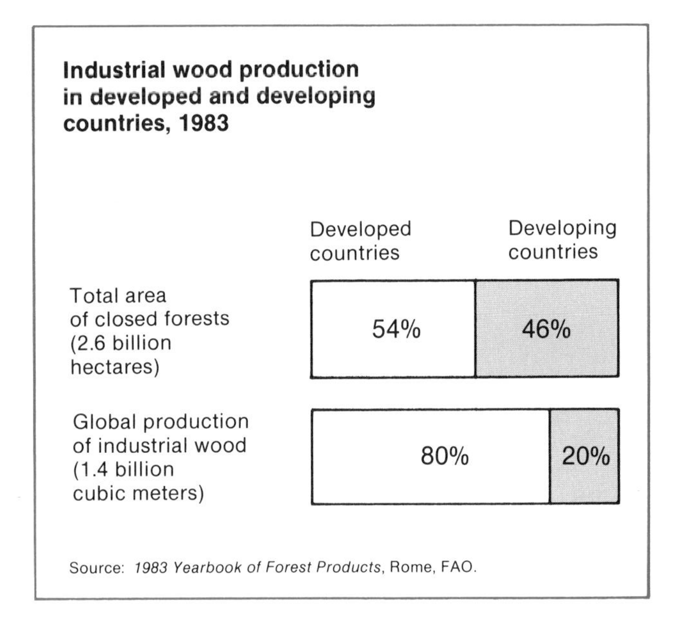

Source: *1983 Yearbook of Forest Products*, Rome, FAO.

Many developing countries have both large natural forests and ecological conditions that are suitable for fast-growing industrial plantations. However, a decline in the area of accessible commercial forests is causing serious problems. In most of these countries current levels of forest management and reforestation fall far short of what is needed to limit imports and sustain exports. Exports to industrialized countries are very important for some developing countries, but the most critical problem for the future is the growing inability of many countries to meet their domestic needs for industrial forest products.

Rising forest product imports

Imports of forest products by developing countries are increasing sharply, even in such countries as Nigeria, Thailand, and Mexico which should readily be able to supply their own domestic needs. Imports have risen from about US$6 billion in the early 1970s to almost US$10 billion today.

In Mexico, the annual value of forest product imports exceeds US$600 million, even though the country has enough forests to be self-sufficient in industrial forest products. Nigeria, once a significant exporter of timber, now imports industrial forest products at a cost of more than US$210 million annually, which nearly equals the value of the 2.5 million tons of food grains currently imported.

In 14 developing countries with suitable conditions for an expanded industrial forestry program, imports of manufactured forest products now total nearly US$4 billion a year. Unless action is taken, this level of imports will continue to rise sharply.

Worsening export outlook

Over the past decade, exports of industrial forest products by developing countries have averaged about US$7 billion (1984 prices) and rank fifth overall in non-oil exports. The value of exports has risen sharply during this time, but it is doubtful that this rate of growth will continue unless additional investment is made to sustain the productivity of industrial forest resources.

In a number of countries, notably Cameroon, Gabon, Ivory Coast, Malaysia, and the Philippines, current rates of timber harvesting and insufficient investment in forest management and reforestation will lead to a sharp decline in exports within 10-15 years. On a smaller scale, the same trend is perceivable in many other countries. In Ghana, for example, exports have fallen from a high of 124 million cubic meters in 1973 to 11 million cubic meters in 1982.

By the end of the century, the 33 developing countries that are now net exporters of forest products will be reduced to less than 10, and total developing country exports of industrial forest products are predicted to drop from their current level of more than US$7 billion to less than US$2 billion.

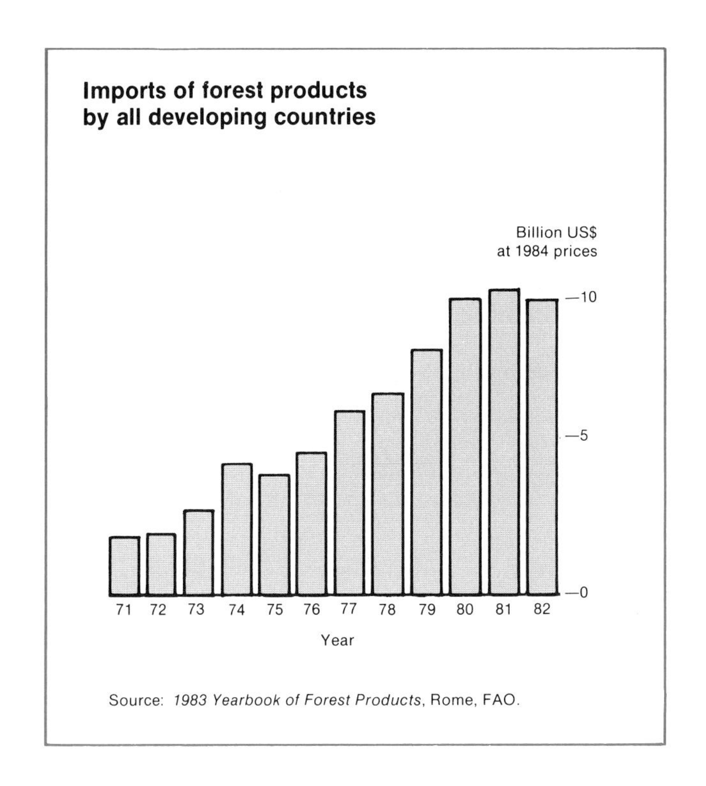
Imports of forest products
by all developing countries
Billion US$
at 1984 prices
—10
—5
—0
71 72 73 74 75 76 77 78 79 80 81 82
Year
Source: 1983 Yearbook of Forest Products, Rome, FAO.

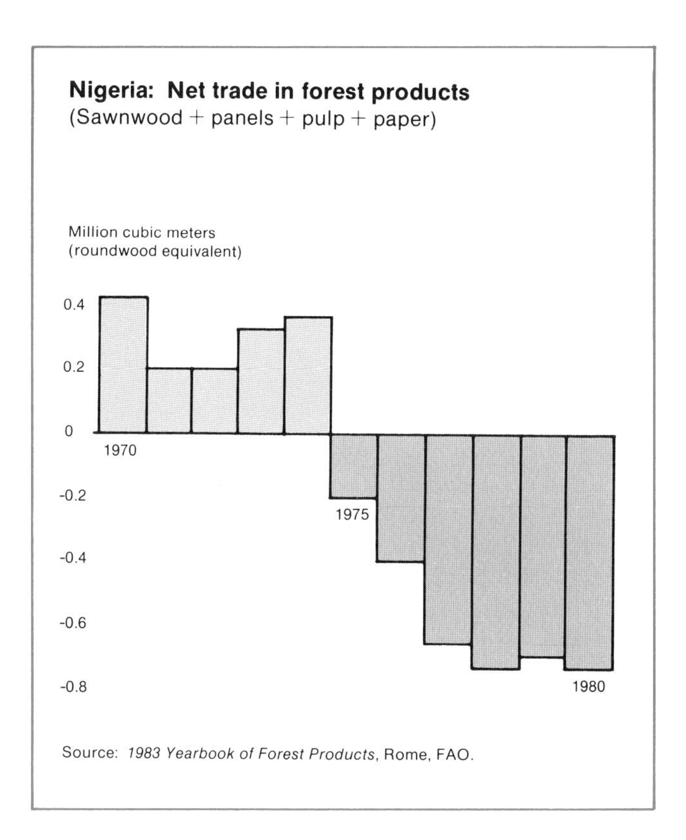
Nigeria: Net trade in forest products
(Sawnwood + panels + pulp + paper)
Million cubic meters
(roundwood equivalent)
0.4
0.2
0
-0.2
-0.4
-0.6
-0.8
1970
1975
1980
Source: 1983 Yearbook of Forest Products, Rome, FAO.

Threatened tropical rain forests

Tropical rain forests are the most biologically diverse ecosystems on earth. They are estimated to contain almost half of all known animal and plant species. However, most tropical species, especially the insects, have not yet been described or cataloged by scientists. If destruction of tropical rain forests—which account for 60% of the world's annual loss of forests—continues unabated, an estimated 10-20% of the earth's biota will be gone by the year 2000.

Dr. Alwyn Gentry examines a plant in one of the earth's least studied rain forests—the Chocó region of Colombia. Deforestation is destroying untold numbers of tropical plant and animal species whose potential benefits are still unknown. (Photo by James P. Blair © 1983 National Geographic Society.)

A single hectare of Amazon rain forest has been known to contain up to 230 tree species, compared with the 10 to 15 species normally found in a hectare of temperate forest. The forests of Borneo have a similarly high diversity of tree species. In Madagascar, only 10% of the original forest remains, yet these forests harbor an extraordinarily large number of endemic plants and animals. For example, more than 20 species of primates are found there and nowhere else. Further loss of Madagascar's remnant patches of tropical forest inevitably will jeopardize many of its endemic species.

Tropical forests yield a wide array of useful products such as essential oils, gums, latexes, resins, tannins, steroids, waxes, edible oils, rattans, bamboo, flavorings, spices, pesticides, and dyestuffs. Many of these materials never enter the commercial market; they are gathered free by local people and are essential to their well-being. But they are also the origin of myriad products manufactured and consumed daily in developed countries, including foods, polishes, insecticides, cosmetics, and medicines. In Indonesia alone, rattans (climbing palms used for cane furniture, baskets, and matting) generate substantial export income, and global trade in rattan end-products totals US$4 billion.

More than 50% of modern medicines come from the natural world, many of these from tropical plants. Two anticancer compounds, for example, derive from the periwinkle plant found only in Madagascar. With these anticancer drugs, there is now a 99% chance of remission in children suffering from lymphocytic leukemia and a 58% chance of remission from Hodgkin's disease. Synthesis of many naturally derived drugs is not commercially feasible, and even for drugs that can be synthesized, the chemical blueprints provided by wild plants are often needed. In developing countries, where modern medicines are often unavailable or too expensive, naturally derived medicines from undisturbed tropical forests may be the primary source of health care.

The centers of origin of many food plants are in the tropics. As tropical lands are converted for human use, ancestral stocks of these plants are jeopardized or lost. Only three species—corn, wheat, and rice—produce two-thirds of the world grain crop. The food supply of the entire world depends on maintaining plant resistance to pests and disease, and resistance is often restored or maintained by cross-breeding with wild populations of the same species. Several wild and domesticated food plant varieties have become extinct and many more are seriously threatened. The gene for semidwarfism that improved production in Asian rice came from a primitive Taiwanese cultivar. Resistance to virus came from a different wild donor species, one that probably evolved in the Silent Valley, a seriously threatened region of India. Incorporation of this gene into new rice varieties has greatly benefited people who depend on this major world crop.

Despite the many uses of tropical plants, less than 1% of them have been screened for their potentially useful properties. Further degradation of large areas of tropical forests will deprive future generations of the chance to retain and broaden the genetic base for food crops, medicines, and other useful products.

Deforestation can be arrested

Solutions are known

The prognosis for tropical forests is indeed grim if action is not taken soon. However, there is still a strong basis for hope. Deforestation can be arrested and ultimately reversed. Although there have been many failures, decades of experience have demonstrated successful solutions to deforestation and land misuse. However, these efforts have been isolated and on far too small a scale to address the problem.

Based on lessons learned from both successful and unsuccessful experiences in the past, enough is known to launch a concerted effort on a broad front to combat deforestation. However, the scale of action required cannot possibly be achieved by government foresters alone. An "across-the-board" effort, involving both the public and private sectors, from government ministries to local community groups, is needed in order to rapidly expand tropical forest conservation and development programs.

Governments must take the lead

Success in reversing deforestation will depend on political leadership and appropriate policy changes by developing country governments to support community-level initiatives. Short-term measures will not solve the problem. Neither will narrowly focused action within the forestry sector. A sustained commitment to forestry, agriculture, energy, and related rural development programs is required.

Solutions outside forestry are essential

Because some policies and practices in agriculture, energy, and other sectors lead to forest destruction, many of the solutions to deforestation must come from outside the forestry sector. Priorities include—

- More intensive agriculture and rural development programs to help the 250 million people already living within tropical moist forests establish sustainable farming systems that do not destroy the forest and to help settle the millions of people living adjacent to threatened forests to minimize further encroachment
- Accelerated land reform programs and expanded employment opportunities to provide some of the developing world's smallholders and landless people with alternatives to forest destruction
- Greater efforts and political commitment to channel future agricultural settlement into nonforest areas and into already deforested areas suitable for agriculture
- Integrated land-use planning that optimizes use of land for agriculture, forestry, conservation, and other productive activities on a sustainable basis, while minimizing the negative impacts of transportation, irrigation, and resettlement schemes on tropical forest ecosystems
- Research to develop sustainable farming systems that combine trees and food crops on the millions of hectares of marginal lands or wastelands
- Revision of government fiscal policies outside the forestry sector (such as subsidies for large-scale cattle ranching) that encourage exploitation, depletion, or waste of forest resources to a greater extent than could be economically justified or commercially profitable without government intervention.

The changing role of foresters

To support the changing emphasis in developing countries from industrial to farm and community forestry, foresters and forestry agencies must make some radical changes in their own policies, priorities, and practices.

In particular, foresters need to—

- Establish policies that encourage local involvement in forestry activities and work more closely with people at the local level by involving them in identifying, planning, and implementing forest protection and management activities.
- Expand mass-media publicity and extension support for forestry conservation and development on farmlands and wastelands outside government-controlled forest reserves. Through education, extension, and awareness programs, encourage recognition of trees and forests as worthwhile "crops" to be cared for in their own right.
- Decentralize tree seedling production and other forestry operations and involve individuals more directly in these activities through local community groups, nongovernmental organizations, and schools.
- Give more attention to conservation programs that can help to increase protection of and research on tropical rain forests.
- Use lower cost technologies such as direct seeding and more intensive mass-production techniques to accelerate tree planting programs.
- Place greater emphasis on multipurpose trees to provide people with timber, poles, fuelwood, fruit, fodder, fiber, and other nonwood forest products.
- Intensify research on agroforestry, management of secondary or degraded forests, and ways to involve local people in forestry.
- Modify and expand forestry training and education programs to place greater emphasis on extension skills, agroforestry, and conservation of forest ecosystems.
- Refrain from converting natural forests to plantations when other suitable land is available.
- Revise government fiscal policies in forestry, such as lenient forest concession agreements, to encourage sustainable management of natural forests and plantations.
- Quantify more precisely the negative effects of deforestation on agricultural productivity, employment, rural incomes, and the balance of trade.
- Work more closely with planners in agriculture, energy, industry, and other sectors to design broadly based agriculture and energy programs in which forestry will play a vital, though not always the lead, role.

Local participation determines success

As important as political leadership is to a successful action program, the key ingredient is active participation by the millions of small farmers and landless people who daily use forests and trees to meet their needs. Countless rural development projects have failed to make a long-term impact because of inadequate involvement of local people. Greater attention must be given to creating incentives for local participation and ensuring that communities are involved meaningfully in project planning and implementation. The roles of women and nongovernmental organizations are especially important.

Creating incentives

Governments need to establish policies that encourage local participation in rural tree planting programs and natural forest management. Forestry codes and laws affecting land and tree tenure; prices for poles, fuelwood, fruit, and other forest products; and the cost and availability of seeds and seedlings of desired species need to be reassessed as potential incentives or disincentives to participation.

Incentives must also be incorporated into development project design. People will not participate in tree planting or related activities if they do not perceive it to be in their interest. Project design must be based on sufficient knowledge of local social, cultural, and ecological conditions as well as of people's perceptions and attitudes. Local participants in a project must be assured of reaping the benefits of their labor.

Involving women

Women play important and in many regions dominant roles in food and livestock production and in the use and management of trees. An increasingly apparent trend in rural areas is the rise in the number of woman-headed households. As a result, women are assuming new roles and responsibilities.

Women and children often suffer disproportionately from deforestation and its aftermath. Women generally are responsible for collecting fuelwood. As fuelwood becomes scarce, they must spend more and more time gathering it and are thus diverted from other household, childcare, or revenue-earning tasks.

Lesotho women working on a tree planting project.

Despite the important economic and social roles of women, forestry and other rural development projects continue to be designed without adequately considering their effect on women or the role of women in their implementation. Although their role in development projects is often overlooked, women have made important contributions. For example, they have carried out soil conservation measures (Lesotho), planted trees (China, El Salvador, Honduras), done nursery work (India), introduced and promoted fuel-efficient stoves (Honduras), and led conservation movements (the Chipko movement in India).

Better information on women's work patterns, their role in the community, and their perceptions of problems and solutions is needed. This requires more involvement of women in extension work. Women must also be represented at the professional level in program planning and project design.

Nepalese villagers discuss a tree planting project.

Nongovernmental organizations: A bridge to the local level

Special attention also needs to be paid to the role of nongovernmental organizations (NGOs) in managing natural tropical forests and in tree planting. An estimated 5,000 NGOs are involved in forestry worldwide, and hundreds of organizations aim, as their primary purpose, to protect forests or to rehabilitate degraded areas.

By working at the local level, often over a long period and with small amounts of money, NGOs can do much to stimulate community involvement in forestry. NGOs often can act as intermediaries between government bureaucracies and local people, and many projects are carried out by NGOs, often with major funding from the development assistance agencies. The role of NGOs in forestry is expanding rapidly, and their involvement will be a vital ingredient in overcoming forestry problems in most developing countries.

Development assistance agencies can do more

Development assistance to forestry is small, particularly in relation to the magnitude of the problems. Worse, it is declining relative to other sectors. The World Bank and the Inter-American, Asian, and African Development Banks allocate less than 1% of their annual financing to forestry, the U.N. Development Programme (UNDP) only 2%. Forestry's many contributions to development are unrecognized because forestry activities do not always bring short-term political or economic gains. Short-term efforts to expand agricultural production usually take precedence.

Rural development experience has shown that how money is spent is even more important than how much. The greater emphasis on farm and community forestry and watershed management requires new approaches to project planning and operations. These projects, which involve changing people's land-use practices, require local participation in their design and implementation. This has several important implications for the development assistance agencies.

Greater attention must be given to human and social factors. Information on local social and cultural conditions should be systematically collected and incorporated into project design. Project planning and implementation should be more flexible and emphasize a "bottom-up" approach. Decisionmaking should be decentralized as much as possible. Strengthening the capabilities of national forestry and related institutions, particularly in working at the local level, must be a major investment priority. Often, small amounts of funding are needed over long periods for this type of activity.

Better coordination among development assistance agencies and within single agencies is needed to avoid duplication of effort, working at cross-purposes, or burdening developing countries with funding and administrative demands that exceed their absorptive capacity. For example, infrastructure development (such as transportation and irrigation schemes) and resettlement schemes must be planned and coordinated to avoid wasting or destroying forest resources, jeopardizing forest conservation areas, or making accessible to settlers forest areas that are unsuited for agriculture in the long term.

Average annual lending (1980–84) by all development banks*

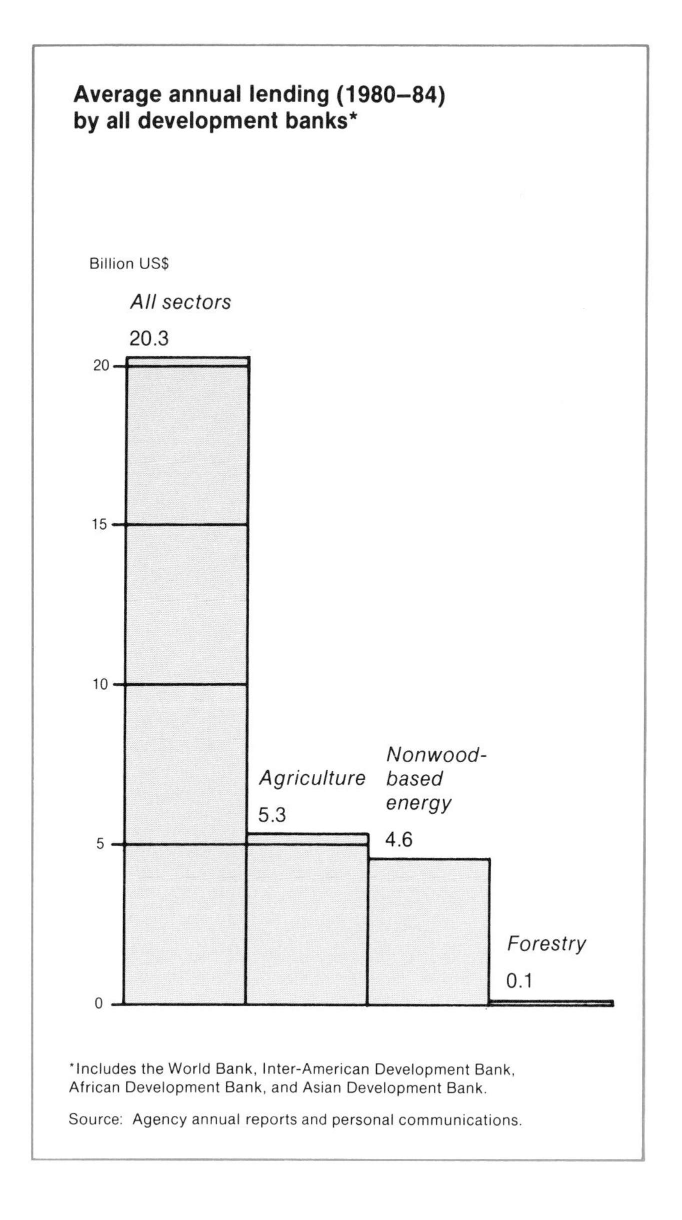

*Includes the World Bank, Inter-American Development Bank, African Development Bank, and Asian Development Bank.

Source: Agency annual reports and personal communications.

The most successful forestry projects have been those where a combination of national government effort and political commitment, assisted by external aid, has created an investment climate that triggered a spontaneous response from local farmers, communities, and the private sector for large-scale self-sustaining programs.

An Agenda for Action

Planning a five-year action program

The Task Force has developed this action program to address deforestation issues on a broad front. In preparing the program, these guidelines were used:

- High priority countries would be identified based on previous studies by the U.N. Food and Agriculture Organization, leading multilateral and bilateral aid agencies, and such nongovernmental organizations as the International Union for Conservation of Nature and Natural Resources.
- Priorities for action should accord closely with those identified as important by national governments and the development assistance agencies. Accordingly, the five priority areas for action identified in FAO's Tropical Forest Action Programme and endorsed by the FAO Committee on Forest Development in the Tropics were adopted.
- The action proposals should be based on successful and well documented development project experiences and also take into account lessons learned from past failures.
- Special attention was paid to identifying small-scale development projects that have potential for widespread replicability.
- Past projects were examined in terms of their success in involving local people. Special emphasis was given to projects with high economic rates of return that have the potential to be self-sustaining.
- Investment needs were treated in the broadest sense to include support for institution-strengthening activities, including research, training, and extension, which experience has shown account for 15 to 25% of total investment requirements.
- In addition to examining investment needs in the forestry sector, an attempt was made to estimate investment needs for agricultural activities that form an integral component of the action plan.
- In estimating investment needs for the period 1987–91, special account was taken of a country's capacity to absorb new investment. The Task Force perceives this as a key factor in limiting accelerated investment in the short term.

The five-year action program proposed by the Task Force is described on the following pages under these five headings:

Fuelwood and agroforestry

Land use on upland watersheds

Forest management for industrial uses

Conservation of tropical forest ecosystems

Strengthening institutions for research, training, and extension.

Each of the five sections opens with a brief summary of the problem and its underlying causes. This is followed by a review of successful experiences that illustrate the range of known solutions. Major policy issues and constraints specific to each section that need to be addressed are identified. Each section concludes with a summary of recommended actions and investments for a five-year action program.

More detailed case studies of successful experiences and country investment profiles can be found in Parts II and III respectively.

Fuelwood and agroforestry

The problem

The rising demand for fuelwood and poles, tree fodder, and agricultural land has greatly accelerated deforestation, bringing in its wake not only shortages of the most important source of household energy in the developing countries but a disastrous series of food crises.

- 63 out of 95 developing countries are faced with inadequate supplies of fuelwood. More than half of the most severely affected countries are in sub-Saharan Africa.
- 34 of the countries with fuelwood shortages have no proven oil or gas reserves, which, combined with low GNP per capita and low rates of economic growth, severely constrains their ability to switch from traditional biomass fuels to fossil fuels.
- Shortages of fuelwood are most acute in semiarid and mountain areas where the productivity of natural woodlands is lowest and the risk of overexploiting the environment is greatest. Fuelwood deficits are increasingly common in densely settled lowlands and in areas of rapidly growing populations and agriculture.

At present rates of consumption, between 1980 and 2000 the annual fuelwood deficit in developing countries will grow from 407 to 925 million cubic meters. This shortfall, which is now met by overcutting forests, is equivalent to the annual output of wood from 80 million hectares of fuelwood plantations.[1] The current rate of tree planting in tropical countries is estimated to be 1 million hectares per year, or little more than 1% of what is required.[2] However, official statistics on tree planting often underestimate the number of trees planted outside of government-sponsored programs.

Several factors in addition to rising fuelwood demand and clearing of forests for agriculture are contributing to wood scarcity:

- Many rural people do not perceive a fuelwood crisis because they have access to freely available alternatives such as animal dung and crop residues.
- In many countries, disincentives exist that seriously constrain tree planting, such as low fuelwood prices and land and tree tenure problems.
- In many areas where there is open access to wood resources, traditional communal systems of management are breaking down.

Countries with major shortages of fuelwood

Region	*Countries affected by acute scarcity of fuelwood or deficits*	*Other countries with areas of fuelwood deficits*
Africa:	Botswana Burkina Faso Burundi Cape Verde Chad Comoros Djibouti Ethiopia Kenya Lesotho Malawi Mali Mauritania Mauritius Namibia Niger Réunion Rwanda Senegal Somalia Sudan	Angola Benin Cameroon Congo Gambia Guinea Madagascar Mozambique Nigeria Swaziland Tanzania Togo Uganda Zaire Zambia Zimbabwe
Asia:	Afghanistan China India Nepal Pakistan Turkey	Bangladesh Indonesia (Java) Philippines Sri Lanka Thailand Viet Nam
Latin America:	El Salvador Haiti Bolivia Peru	Brazil Costa Rica Chile Colombia Cuba Dominican Republic Ecuador Guatemala Jamaica Mexico Trinidad and Tobago

Source: Based on the 1980 study by FAO of fuelwood supplies in developing countries. Fuelwood scarcity was defined as an inability to meet minimum requirements, even by overexploitation of remaining woodlands. Fuelwood deficits indicate that demand is met by harvesting wood faster than it is being replenished.

[1] Assuming an average annual yield of 5 cubic meters per hectare.

[2] Includes fuelwood produced from nonindustrial and industrial plantations.

Success stories

Fuelwood conservation and the use of substitute fuels

- Improved wood stoves made from a clay and sand mixture have been developed successfully in Senegal as part of a national program launched by the Center for Study and Research on Renewable Energy. The most popular type of stove, the Louga model, was designed and is built principally by women. About 3500 stoves of this type were built in the first 2 years of the pilot program, and 77% were in regular use when the program was later evaluated.
- As part of a community forestry project in Nepal, improved cooking stoves have been introduced to decrease per capita fuelwood consumption. More than 700 stoves have been installed, resulting in an average fuel saving of 38%.
- As part of a program aimed at reducing fuelwood consumption and accelerating reforestation, the Forest Research Institute of Korea developed a more efficient system of underfloor heating that is capable of reducing wood use for heating by 30%. Sale of fuelwood to city residents was prohibited to discourage illegal cutting in rural areas. At the same time, the government pushed ahead with rural electrification. As a result of increased use of electricity and more efficient use of fuelwood, the share of fuelwood in total energy consumption fell from 55% in 1966 to 19% in 1979.

Improved management of forests and more efficient conversion of wood into charcoal

- A degraded forest in Ghana was cleared and replanted for the production of pulpwood. Instead of burning the cleared vegetation, which was the standard practice, the waste wood was salvaged and either sold directly as sawtimber or fuelwood or efficiently converted to charcoal and then sold. Food crops were planted for 3 to 4 years between the rows of tree seedlings to help suppress weeds. Returns from the sale of the felled wood and the crops exceeded conversion costs by several hundred dollars a hectare.
- In Uganda, logging residues once wasted have been recovered for charcoal manufacture, increasing the output of charcoal from 200 to 63,700 tons per year.
- In Niger, Senegal, Ivory Coast, and Chad, the productivity of vegetation has been increased on an experimental basis by 20-100% through various combinations of controlled grazing, regulated fuelwood harvesting, and protection from late annual bush fires.

Waste wood from clearing forest that was once burned is now used for making charcoal in the Subri River Forest Reserve in Ghana.

Reforestation through agroforestry, cash crop tree farming, and farm forestry

In many countries it has become clear that progress in agriculture depends on controlling deforestation, regenerating the vegetative cover, and adopting farming practices that compensate for shortened fallow periods. Agroforestry techniques can restore the tree cover and enhance net output of annual crops, livestock, and a variety of perennial tree crops.

Several reforestation programs show that governments can mobilize widespread support for tree planting and organize large-scale seedling production and tree planting programs. In parts of India, Nepal, Kenya, and elsewhere, it has been possible to build up rural forestry extension programs and motivate farmers to plant trees on their own land. People in these countries have demonstrated their responsiveness to growing urban demands for wood and to increasing scarcity and higher prices for fuelwood and poles, by eagerly participating in programs offering seedlings of fast-growing, multipurpose trees. However, experience has shown that it is easier to motivate people to plant trees than to divert time from other activities to tend and protect the trees. Many planting programs have had very low tree survival rates for this and other reasons.

Decentralized distribution of seedlings is an important incentive for tree planting.

Watering seedlings in a tree nursery as part of a reforestation project in Thailand.

• An agroforestry project in Haiti has proven the effectiveness of using existing networks of nongovernmental organizations to promote tree planting in rural areas. Farmers have been quick to respond to the provision of subsidized seedlings and extension services, because they readily appreciate the income-earning potential of growing trees to produce charcoal for urban markets. In less than 3 years, the project has reached thousands of farm families who have planted more than 10 million seedlings, or double the amount planned.

• In Gujarat State, India, a social forestry project was started with the goal of distributing 30 million seedlings annually. Within 3 years, seedling distribution had increased from 17 million to nearly 200 million plants per year, as farmers recognized the income-generating potential of growing trees to meet the market demand for poles. At first, the density of nurseries had been one per 83 villages. Encouragement of nurseries tended by schools and individual farmers increased the density of nurseries to more than one per 10 villages. By 1983, more than 150,000 hectares had been planted in an effort that involved one in every 10 farmers in Gujarat.

• In Uttar Pradesh, India, another social forestry project had a 5-year goal of planting 8000 hectares of woodlots and 22,000 hectares of strip plantations, distributing 8 million seedlings for farm forestry plantings, and rehabilitating 13,000 hectares of degraded forest. The targets were exceeded within 3 years, and in the process 17 million workdays of employment were generated, including 4 million days for women.

• A recent survey in the Kakamega district of Kenya reveals that 72% of farmers have planted trees and 38% are raising seedlings. Trees are planted for fruit, shade, ornamental plantings, boundary plantings, and windbreaks, as well as for fuelwood.

• Agroforestry studies in Senegal have shown that yields of millet and sorghum were 500-1000 kilograms per hectare higher in fields with *Acacia albida* trees than in open, treeless fields.

• Maize yields in Nigeria declined to 500 kilograms per hectare under continuous cultivation for 6 years in a control plot, but they were sustained at 2000 kilograms per hectare by mulching with leaves from the leucaena tree. In other "alleycropping" trials, crop yields were 18% higher in plots in which grazing and maize cultivation were rotated between the rows of closely spaced, leguminous trees.

• Growing trees for fuelwood and other uses can be designed to have a positive effect on food production. In the Majjia valley of Niger, millet yields were increased 20–25% by planting windbreaks.

• In China, 30 million hectares of trees have been planted since 1949 to form shelterbelts around farmfields, soil conservation plantations, dune reclamation works, and roadside plantings. The total area planted in China is more than double the combined area of fuelwood and industrial plantations established during the same period in all other developing countries. This extraordinary achievement is largely a reflection of government commitment to the program and the massive mobilization that was organized as a result.

Proposals for accelerated action

General strategy and policy issues

The recommended program addresses the need to manage demand and to increase supplies of fuelwood. Because much of the fuelwood needed in developing countries will have to be grown in rural areas, agroforestry has a major role to play. In addition to increasing fuelwood supplies, agroforestry will have a very positive effect on both food production and rural economies.

The main components of the program are identified in the box.

To apply this strategy, these important policy issues need to be addressed:

Fuelwood pricing

The low fuelwood prices that prevail in rural areas of developing countries reflect the fact that wood is still "freely available," even if it is collected by overcutting. Low prices constrain investment by farmers in growing trees. Incentives such as subsidized seedling distribution may be used to trigger widespread interest in reforestation. More research is needed on the economic justification for such subsidies and their potential role in helping to reduce deforestation.

Distribution of benefits

Community approaches to reforestation and natural forest management have proved difficult, particularly in semiarid environments. A major difficulty is ensuring that the benefits of tree planting and improved management are distributed equitably. In particular, local people who carry out the project must be assured of receiving the benefits of their work.

Efforts organized at the village level are also constrained by the heterogeneity of local communities and the need to reconcile conflicting interests of herdsmen and farmers. Conflicts also arise between rural communities and urban-based woodcutters and charcoal producers.

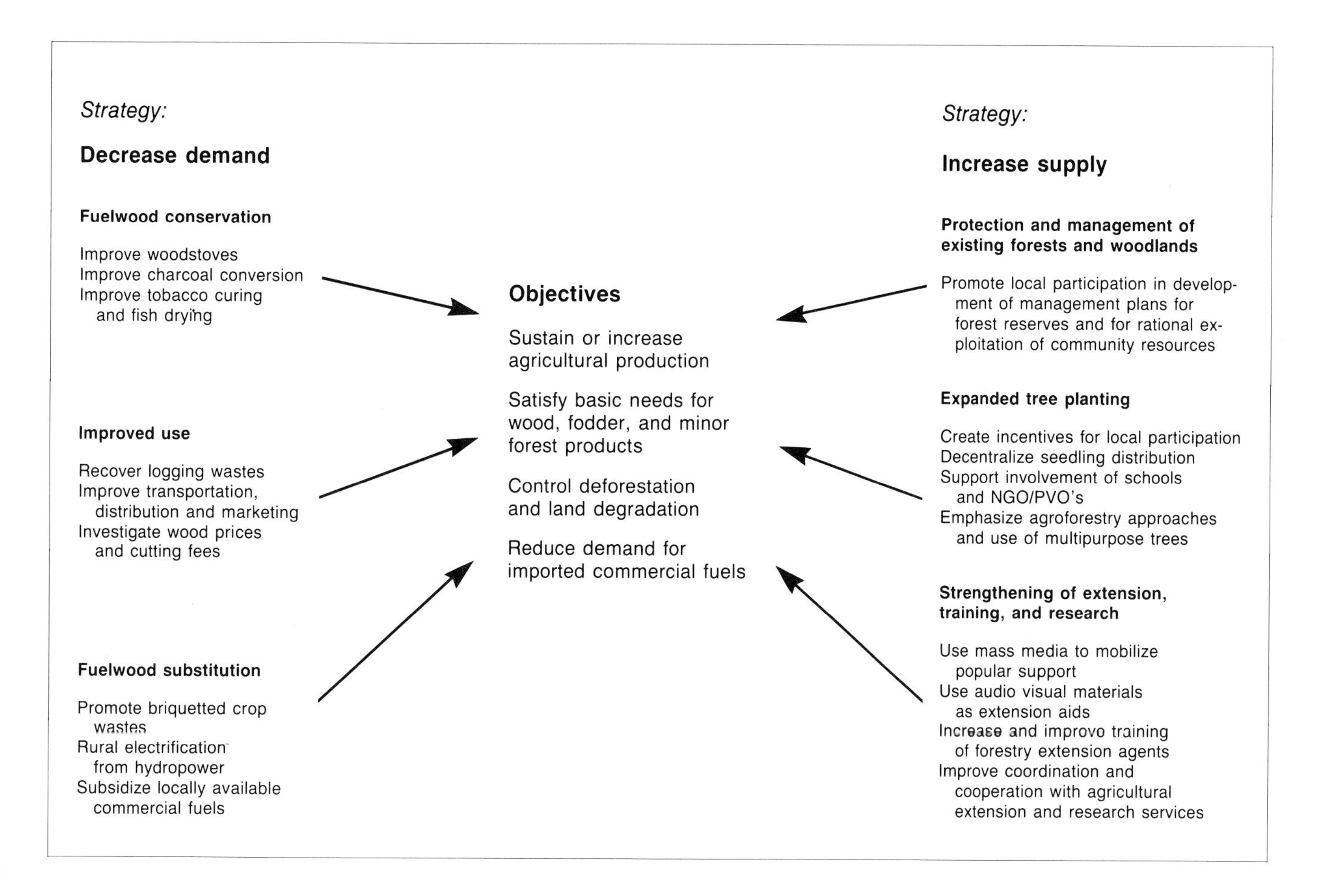

Improved development planning

Because clearing of land for agriculture is the leading direct cause of deforestation, agricultural resettlement and livestock development policies need to be reassessed to ensure they promote sustainable development. The potential for integrating fuelwood production, agroforestry, and soil and water conservation techniques into farming systems needs to be systematically reviewed and carried out.

Recommended actions for a five-year program, 1987–91

- Support programs to develop more efficient cookstoves, building on the experiences gained in Senegal, Nepal, Burkina Faso, Guatemala, and elsewhere. Focus on regions most affected by fuelwood scarcity and price increases.
- Promote more efficient wood use and charcoal production technologies, such as improved recovery of logging wastes and briquetting of crop residues from large-scale farms. Support outreach programs with charcoal producers and industrial users of fuelwood (for tobacco drying, brick kilns, bakeries) to increase the efficiency of fuelwood use.
- Evaluate the potential impact and economic rationale for encouraging the use of locally available commercial fuels or renewable biomass substitutes for household energy sources.
- Consider changes in woodcutting fees and current methods of collection that would give rural communities greater control over tree harvesting and eliminate "free" wood. Reassess wood prices and wood distribution networks; evaluate the potential for increasing farmgate prices of wood and for lowering in cities the retail prices of wood produced by private and government plantations.
- Increase local involvement in protecting and managing forests and woodlands. Productivity can be increased by organizing community control over access and use and by using techniques that conserve soil moisture and enhance natural regeneration.
- Strengthen forestry extension activities by training more agents and developing audiovisual materials that clearly demonstrate the relationships between the managed use and sustained productivity of water, soils, pastures, and forests, as was done in Nepal and Burkina Faso.
- Promote tree planting around family homesteads and cultivated areas by removing legal and other constraints and disincentives to on-farm tree planting.
- Develop low-cost technologies (e.g., direct seeding or the "basket" system) for seedling distribution, and promote decentralized seedling production involving schools and nongovernmental organizations, as in Haiti and India.
- Encourage local support and participation in tree planting by increasing the use of fast-growing, multipurpose tree species that meet people's perceived needs. Monitoring and detailed surveys, such as those underway in Malawi, Kenya, and India, are needed.
- Encourage private sector involvement in establishing and maintaining plantations; give priority to planting deforested lands not suited for annual crops.

Investment needs

The proposals cover 32 countries in which large populations are affected by acute fuelwood shortages, and in which agroforestry could play an important role in increasing fuelwood supplies and food production. This is only a listing of the countries where the problem has already been well documented; other countries will need similar support.

- The estimated cost of a 5-year investment program in these 32 countries is US$1.9 billion. Much of the proposed funding would be absorbed by several larger countries (China, India, and Brazil) which have already made considerable progress in developing the institutions for expanded reforestation and wood energy programs.
- For most of the other countries included in this analysis, annual investment averages US$5 million per country and represents a doubling of current assistance to the forestry sector.
- The proposed program includes a marked increase in support for fuelwood conservation, improved management of existing woodlands, and institutional strengthening to support field activities and promote policy changes.
- Investment in forestry research, training, and extension accounts for 20% of the investment needs.

Fuelwood and agroforestry

Summary of needed investments, 1987–91

Africa	*Million US$*
Botswana	15
Burkina Faso	25
Burundi	20
Cape Verde	15
Chad	14
Ethiopia	40
Kenya	48
Lesotho	10
Madagascar	30
Malawi	24
Mali	30
Mauritania	16
Niger	20
Nigeria	50
Rwanda	30
Senegal	25
Somalia	15
Sudan	35
Tanzania	30
Uganda	15
Asia	
Bangladesh	52
China	250
India	500
Nepal	30
Pakistan	40
Sri Lanka	30
Latin America	
Bolivia	25
Brazil	400
Costa Rica	15
El Salvador	10
Haiti	15
Peru	25
Total (32 countries)	1899

Land use on upland watersheds

The problem

Destructive land use on upland watersheds is taking place on a vast scale throughout the developing world. The degradation of upland stream source areas results in massive soil erosion and sedimentation of rivers, dams, and reservoirs. The consequences in the lowlands are increased severity of flooding which leads to losses of crops, land, buildings, and even human life; disruption of irrigation systems and reduced crop yields; and decreased power generation for urban areas. The problem of watershed degradation is complicated because downstream impacts can occur far away from the source of damage and across political boundaries.

An estimated 160 million hectares of upland watersheds have been seriously degraded in Africa, Asia, and Latin America. Countries with serious upland watershed problems include—

Africa	*Asia*	*Latin America*
Burundi	China	Argentina
Ethiopia	India	Bolivia
Guinea	Indonesia	Brazil
Kenya	Nepal	Chile
Lesotho	Pakistan	Colombia
Madagascar	Philippines	Costa Rica
Mozambique	Thailand	Ecuador
Tanzania		El Salvador
Uganda		Guatemala
Zimbabwe		Haiti
		Jamaica
		Mexico
		Nicaragua
		Panama
		Peru

Land misuse in rural areas stems in great part from increasing population pressures and political and administrative neglect by urban-based government authorities. Other major causes of watershed degradation are described below.

High-rainfall uplands

Rapid population growth and the search for food, fuel, and fodder have caused invasion and destruction of upland forests. Subsequent misuse of exposed slopes through intensive overgrazing by free-ranging livestock and unprotected cultivation is causing widespread degradation of soil and water resources.

Medium-to-low-rainfall uplands

Because of their harsher environment, particularly low and erratic rainfall, medium-to-low-rainfall uplands have lower carrying capacities for people and livestock. When undisturbed, these areas develop a continuous tree, shrub, and grass cover that protects the soil. This cover is vulnerable to misuse and at low rainfall levels it recovers slowly. Rising human and livestock populations are reducing these areas to unproductive wastelands on a vast scale by fuelwood gathering, uncontrolled and unproductive overgrazing, and depletion of soil fertility through continuous cropping.

Success stories

High-rainfall uplands

• In Nepal, increased fodder and fuelwood supplies from planting grasses and multipurpose trees, and tethering or stall-feeding livestock, have provided enough manure for a second grain crop each year and quadrupled family incomes. The key incentive was to offer wages for tree planting for the first 2 years only.

Stall-feeding of livestock reduces grazing pressure and allows for revegetation of denuded slopes.

• In India, reforestation and soil conservation measures have stabilized watersheds above major dams. In the Damodar Valley, floods have been moderated, the rate of sedimentation has been reduced, more water is available for human and livestock use, and crop yields have increased.

• In tropical southern China, reforestation by manual contour ditching and intensive tree planting has checked erosion and permitted fish farming and small-scale hydropower development. One commune put 10,000 men, women, and children into the field and planted 3.4 million trees in one year. Their income is now twice the average for the province.

• In Colombia, an innovative program to transfer resources from lowland beneficiaries of hydropower development to upland farmers is being started. A sales tax on electric power from major hydroelectric plants will be used to promote proper land use for stabilizing upland watersheds through soil conservation and reforestation.

Medium-to-low-rainfall uplands

• In the Ethiopian Highlands, the largest soil conservation program in Africa is being supported by many multilateral and bilateral donors. More than a million people from 8000 Peasant Associations are carrying out Food for Work programs. Under the program, each worker is provided a family food payment of 3 kilograms of wheat or maize and 120 grams of vegetable oil for each day's work on conservation projects such as terracing of steep slopes and tree planting. Peasant Associations have played a major role in the success of the projects because of their ability to mobilize labor quickly and efficiently.

Members of Peasant Associations in the Ethiopian Highlands are rehabilitating degraded land through food-for-work programs.

In China, watersheds are being rehabilitated on a vast scale through massive deployment of hand labor using conventional soil conservation methods.

• In the low-rainfall loess plateau region of China, abandoned agricultural land destroyed by soil erosion is being successfully reclaimed. On the edges of the wasteland where road access remains, conventional soil conservation methods have been successfully applied by massive deployment of hand labor. In the past 5 years, Chunhua County has planted 15 million trees and sediment transport from the watersheds has already been halved. Bare hills that are inaccessible by road are being rehabilitated by aerial seeding. Some 10,000 hectares have been established successfully in experimental programs using common alfalfa seed at 2 kilograms per hectare.

• In Uganda, severely overgrazed watersheds were equipped to measure floodflow and rainfall penetration into the soil. After 4 years of measurements, excessive grazing was allowed to continue on one watershed while the adjacent watershed was treated by bush clearing and controlled grazing. Simple rotational grazing fattened more livestock on the managed watershed than were starving on the traditionally grazed control valley. Rain penetrated one meter further into the soil and flood peaks were halved on the managed area.

Proposals for accelerated action

General strategy and policy issues

The recommended program addresses the need to change land-use practices to stabilize and rehabilitate degraded upland watersheds. Technically proven and economically sound methods are available for different social and ecological conditions. Experience has shown that proper land use can sustain and improve the productivity of both forests and agricultural land. Solutions vary with local conditions, and pilot projects are needed in countries that are only beginning to address their upland watershed problems. The recommended program focuses on three main activities:

- Establishing tree and grass cover to stabilize upland areas and provide adequate supplies of fuelwood, fodder, and building poles
- Controlling livestock grazing
- Developing sustainable farming systems.

To apply this strategy, these important policy issues need to be addressed:

Interdependence of upland and lowland communities

- Government policy must be based on the interdependence of upland and lowland communities.
- Governments must provide more support for upland communities to deal with problems of poverty and remoteness from technical help, both of which cause land misuse.

Multidisciplinary approaches

- Because several disciplines are needed to rehabilitate and manage degraded watersheds, action cannot be carried out by government foresters alone; they must include agriculturalists, water resource engineers, community administrators, and others.
- In-service training is needed to ensure that sound technology is applied in each of the several disciplines involved.

Incentives

- Changes in land-use practices require incentives and the full participation and support of local communities. The most practical incentives are short-term daily wages for soil conservation and tree and fodder planting. Another important incentive is provision of seed and fertilizer at subsidized prices.
- Hill farmers on land that can sustain agriculture must be assured security of land tenure and protection from invasion by migrants. Farmers on land unsuited for agriculture should be resettled to areas capable of sustaining them.
- Better pay and status for administrative and professional staff in remote rural areas are essential. Career prospects now depend on working in the cities.

Recommended actions for a five-year program 1987–91

The proposals for accelerated action over the next 5 years cover 11 countries that have begun to take corrective measures known to have achieved success. These are some of the countries from which evidence is available because they have seen the need to correct the current misuse of their uplands and have begun practical rehabilitation with the support of external aid. However, in all these cases, the means are not yet adequate to win the race against time to preserve critical upland soil and water resources. The countries are China, Colombia, Ethiopia, India, Indonesia, Kenya, Madagascar, Nepal, Pakistan, the Philippines, and Zimbabwe.

This list will need to be expanded as other countries take up the challenge of their rural watershed problems and seek help in solving them. No developing countries can afford either to ignore the dangers to their soil and water resources as populations increase or to ignore the capacity of their lands to sustain future population growth.

High-rainfall uplands

- Provide incentives for local communities to reforest upper slopes and increase fodder and fuelwood production through tree planting and protection of remaining resources.
- Tether and/or stall-feed livestock.
- As fuelwood supplies increase, maintain soil fertility by applying animal dung and crop residues instead of using them as fuel.
- Reduce soil erosion by simple manual techniques, such as repair and construction of bench terraces and plugging of gulleys with check dams. Plant trees and grasses along the terrace edges or lines.

• Encourage farmers to adopt sustainable agroforestry practices.

• Expand small-scale water resources development. Plastic pipe to carry water from springs is a major incentive to stall-feeding of livestock.

Medium-to-low-rainfall uplands

• Maintain essential vegetation cover on steep slopes; this requires that community needs for both fuelwood and fodder be met by planting trees and grasses.

• Rotate livestock to preserve cover on mild slopes and exclude grazing on steep land.

• Intensify production through agroforestry on the more fertile areas to relieve pressure on marginal lands.

Investment needs

No data for the total area of eroded watersheds are available on a worldwide basis such as that collected for fuelwood by FAO. Both the scale of the damage and the timetable for remedial action are highly specific to the social, political, and economic circumstances of each country. No global total for investment can be estimated with confidence, but the 16 countries listed in the table, for which proposals for action total US$1231 million, represent about two-thirds of the problem.

Land use on upland watersheds

Summary of needed investments, 1987–91

Africa	*Million US$*
Ethiopia	100
Kenya	35
Madagascar	10
Zimbabwe	46
Asia	
China	135
India	500
Indonesia	100
Nepal	15
Pakistan	45
Philippines	120
Latin America	
Brazil	10*
Colombia	50
Ecuador	15*
Jamaica	10*
Panama	20*
Peru	20*
Total (16 countries)	1231

*Preliminary estimate, pending additional research.

Forest management for industrial uses

The problem

Many developing countries have both substantial natural forests and ecological conditions suited for fast-growing industrial plantations. Nevertheless, in most such countries current levels of forest management and reforestation fall far short of what is needed to limit imports and sustain exports.

Consumption of forest products has outstripped increases in domestic production. As a result of inadequate attention to management and development of their industrial forest resource base, more and more developing countries face burdensome bills for importing forest products. In several countries, exports of forest products have increased, but this growth cannot be sustained without increasing investments to maintain supply. The critical countries are—

Africa	*Asia*	*Latin America*
Cameroon	Burma	Argentina
Congo	China	Brazil
Gabon	India	Chile
Ghana	Indonesia	Colombia
Ivory Coast	Malaysia	Costa Rica
Liberia	Pakistan	Ecuador
Nigeria	Papua New Guinea	Guatemala
Swaziland	Philippines	Jamaica
Uganda	Thailand	Mexico
Zaire		Peru
		Venezuela

Several factors have contributed to this situation:

• Every year, 5 million hectares of closed tropical forests are logged. Frequently, only a few of the highest valued and more easily marketed species are extracted from uncut forests. This process disturbs much of the remaining vegetation and reduces significantly the commercial value of the secondary forest that grows back.

• Reforestation has not kept pace with logging and deforestation. Less than 600,000 hectares of industrial plantations are planted each year in developing countries. This compares to the annual logging rate of 5 million hectares and an annual deforestation rate of 11.3 million hectares.

• Industrial forest resources have not been well managed. Over the past 30 years, there has been minimal investment in protecting and intensively managing forests that have been logged. Many plantations are not well maintained, protected from fire, or regularly thinned and harvested. Because of an inability to properly manage existing plantations, yields have sometimes been lower than expected.

• When new roads provide access to forests, uncontrolled encroachment by farmers (and in some areas ranchers) often follows. Each year, more than 7.5 million hectares of closed forests are lost by conversion to agriculture.

• Forest management is also hampered by a shortage of well trained personnel, insuffcent investment in research, and inadequate administrative structures and financing mechanisms.

Success stories

More intensive management of natural forests

Attempts to manage natural forests to reproduce the valuable species removed in a first cut have not been widely successful, but there have been encouraging results in a few locations, including Ghana, India, Congo, Gabon, and Suriname. In Gabon, species like the mahoganies *(Okoumea)* have been naturally regenerated and enrichment planting has increased the value of the forest resource.

Increased use of lesser-known hardwoods

Malaysia has successfully developed local markets for many lesser-known hardwood species once regarded as "weed trees." By 1981, unexploited hardwood species accounted for about 12% of the total log intake of sawmills in Peninsular Malaysia. The intake of these species in plywood/veneer mills was even higher (27%). In Cameroon, Indonesia, Colombia, and the Philippines, previously noncommercial species have also been successfully used for pulpwood.

Fast-growing industrial plantations: Zambia, Chile, and Brazil

The most dramatic successes in industrial timber production are large-scale industrial plantations.

Over the past 20 years, Zambia has established industrial plantations capable of meeting its projected industrial timber needs through the end of the century. Beginning in the mid-1960s, a sustained program of industrial reforestation (primarily pines and eucalypts) surpassed its targets and by 1983 had successfully established 45,000 hectares of plantations. This has helped reduce pressure on the country's diminishing natural forests and developed a highly productive source of roundwood needed in its copper mines and other industries.

Chile has established 1.1 million hectares of pine plantations over the past 20 years through the combined efforts of government and the private sector. Favorable forestry policies, including specially designed tax laws and subsidies, also played a major role in the success of the reforestation programs. By the end of this century,

Chile: Exports of forest products

Source: *1983 Yearbook of Forest Products*, Rome, FAO.

Chile will have the potential to produce nearly as much industrial wood as Finland does today.

Sawmills and pulp and paper mills have been built. Chile is now supplying all of its domestic demand for forest products and is exporting logs, sawnwood, pulp, and paper. The value of forest-based exports has reached US$350 million per year.

Within 20 years, Brazil has established more than 4 million hectares of plantations of both eucalypts and pines, not only meeting a rapidly growing domestic demand but also providing the basis for a profitable export business. Industrial wood production in Brazil rose from 24 million cubic meters in 1973 to more than 58 million cubic meters by 1983.

Small landholder tree planting for cash crop and industrial wood production

Plantations of various fast-growing species have become a profitable cash crop for farmers in a number of developing countries. Poplars are grown as a cash crop in Pakistan, and farmers in India have enthusiastically planted eucalyptus to produce poles for sale in local markets. In the Philippines, a combination of technical and financial assistance to smallholders provided by the Paper Industries Corporation of the Philippines and the Development Bank of the Philippines has led to the participation of more than 3800 farmers in planting *Albizzia falcataria* to produce wood for a pulp and paper mill on the island of Mindanao. A total of 23,000 hectares of tree farms had been established by 1984.

Plantations of valuable hardwoods

Slower growing but higher valued hardwood species such as teak have been planted successfully in many tropical countries in Asia, where teak originates, as well as in Africa and Latin America. These plantations can help replace cutover forest and increase future wood production.

Proposals for accelerated action

General strategy and policy issues

The recommended program focuses on three major types of activities:

Protection and management of natural forests

More intensive use of existing resources, particularly lesser-known hardwood species

Accelerated industrial reforestation.

The potential for greater production is significant. Assuming annual yields of 2 cubic meters per hectare from managed natural forests, existing areas of logged-over forests (about 210 million hectares) could be more intensively managed to supply 85% of developing countries' demand for industrial timber in the year 2000.

Alternatively, assuming average annual growth of 15 cubic meters per hectare in plantations, total demand could be met by converting only 33 million hectares (or about 15%) of the existing logged-over area into fast-growing plantations.

In terms of technical requirements for certain types of wood, the best solution probably is some combination of these two approaches.

To apply this strategy, these important policy issues need to be addressed:

Improved management of natural forests

• Much greater political commitment to the goal of sustainable resource management is needed. Some progress in this regard can be found in Peninsular Malaysia, for example, and in Indonesia.

• Government policies that encourage unsustainable forest exploitation—such as inappropriate forest concession agreements—should be identified and changed.

• Proven techniques for regenerating natural forests after logging need to be more widely disseminated, and economic criteria need to be introduced. Further development of techniques for the use of lesser-known species could greatly improve the economics of natural forest use.

Industrial plantations

• Forest land must be controlled or allocated to (1) protect industrial forest areas from invasion by migrants to allow sound silivicultural management; (2) assure landowners that they will be able to market industrial wood at a reasonable price; and (3) assure wood-consuming industries, particularly those requiring large long-term investment, that they will have access to a steady supply of wood at reasonable prices.

• Financial incentives are needed to encourage investment in reforestation and forest management. The time that it takes to grow industrial timber and the prevailing interest rates in many countries discourage investment in industrial reforestation. Where planting can be justified on economic grounds, governments can help overcome this problem by using fiscal and other incentives. Mechanisms also are needed to ensure higher standards of plantation maintenance, fire protection, exploitation, and regeneration.

• Well conceived research programs are needed to capitalize on the potential for fast-growing industrial plantations and to increase our knowledge of multiple-use management. Better management is needed for closed forests that have been logged and forests on sites that are steep, infertile, inaccessible, or otherwise unsuited for agricultural development. Through improved management, these lands can yield large benefits in watershed management, protection of wildlife habitats, and nonwood as well as wood products.

Recommended actions for a five-year program, 1987–91

Specific recommendations for action have been developed for 28 countries. On the basis of these country-specific analyses, the following actions are recommended:

• Increase the area of managed tropical forest from 34 million hectares to 67 million hectares over the next 5 years by protecting areas that have already been logged and by improving logging practices. Given current manpower constraints in most countries, "management" will usually constitute no more than the development of forest management plans. On this basis, assuming an average annual management cost of US$5 per hectare, the cost would be on the order of US$165 million annually by 1990, or about US$550 million over the 5-year period.

• Increase the rate of industrial plantation establishment to 950,000 hectares annually by 1990. The additional cost, assuming an average cost of US$500 per hectare, would total US$750 million over the next 5 years.

• Establish forestry funds and otherwise assist appropriate administrative and commercial organizations to provide more intensive management of existing and future plantations.

• Greatly accelerate research on regeneration and sustained-yield management of natural and planted forests; research should be better coordinated and assistance should be provided for disseminating the results.

• Increase the efficiency of harvesting and use of tropical timber resources.

Investment needs

The estimated cost of a 5-year investment program for the 28 countries listed in the table is US$1.6 billion. Perhaps half of the needed investment could come from the private sector if favorable tax policies and other incentives were adopted.

Forest management for industrial uses

Summary of needed investments, 1987–91

Africa	*Million US$*
Cameroon	20
Congo	20
Ghana	10
Ivory Coast	75
Liberia	15
Nigeria	35
Uganda	25
Zaire	10
Asia	
Burma	30
China	285
India	190
Indonesia	50
Malaysia	40
Pakistan	20
Papua New Guinea	15
Philippines	40
Thailand	35
Latin America	
Argentina	100
Brazil	325
Chile	50
Colombia	45
Costa Rica	15
Ecuador	20
Guatemala	15
Jamaica	10
Mexico	90
Peru	30
Venezuela	25
Total (28 countries)	1640

Conservation of tropical forest ecosystems

The problem

Destruction or conversion of tropical forests, particularly tropical rain forests, is resulting in widespread disruption of ecosystems and loss of species.

Agricultural clearing by more than 250 million people who live in tropical forests is increasing. Not only has the population of subsistence farmers grown, but in many nations the absolute amount of cleared land available to them has decreased. The concentration of land ownership, which has characterized land tenure in the developing world, has become even more pronounced in recent years. Much of this land is used to produce export products, while in many areas per capita food consumption continues to decline.

Expanding agriculture into tropical forests often is futile because the soils are poor or unsuited to continuous production. This is exacerbated by the shortening of fallow periods because of increasing demands for food. In addition, much good agricultural land now lies fallow, and an even larger amount is managed inefficiently and nonintensively. Improving agricultural efficiency, especially for the small farmer, could greatly reduce pressures on forests.

In Colombia, for example, small farmers produce three times as much food per hectare as do owners of large farms. But because of population pressure and land degradation, these same farmers do not now have enough land to pursue their traditional farming lifestyles, and increasingly they move into urban areas or into the forests of Amazonia, where there are now nine times more people than in the mid-1950s.

Cattle ranching causes widespread loss of tropical forests, particularly in Latin America. Overgrazing degrades pasture and limits forest regeneration. Compared with growing crops, raising livestock is a low productivity use of arable land. Much of the meat goes to the cities or is exported to developed countries, with the income primarily benefiting a small number of large landholders.

Commercial logging affects an estimated 5 million hectares of undisturbed forests each year, and this does not include estimates of illegal logging. Trade records from Thailand and the Philippines, for example, indicate that more trees are logged illegally than legally. Careless logging can lead to ecological damage that is much greater than simply the loss of the logged trees: often 30–60% of residual trees are injured beyond recovery. Large areas are often left bare, leading to soil loss. Logging machinery compacts the soil, reducing water infiltration rates, and increasing soil erosion. Most important, logging roads increase access for farmers who clear additional land for agriculture. Such unintentional opening up of forests occurs worldwide, and it is particularly serious in Amazonia and tropical Asia.

Success stories

The global system of conservation areas

- More than 3000 parks and equivalent reserves covering more than 400 million hectares have been established worldwide. The number of sites has more than doubled over the past 15 years, and many of these areas are in tropical forests. Preservation of wildlife and strict protection of areas such as parks are only one part of the much broader scheme of conservation. New land designations and management methods, allowing a variety of uses and types of exploitation, have been developed. The International Union for Conservation of Nature and Natural Resources (IUCN) is leading a major international effort to determine the gaps in ecosystem coverage in the global system of conservation areas.

Henri Pittier National Park in Venezuela protects tropical rain forest for the future.

• UNESCO's Man and the Biosphere Programme (MAB) promotes the establishment of Biosphere Reserves—multiple-use conservation areas containing both natural land and areas modified by human activity. Undisturbed core areas are managed to maintain biological diversity and ecosystem processes. As environmental monitoring sites, they serve as benchmarks for measuring long-term changes. Other zones of a biosphere reserve are managed to allow a variety of human activities, including farming or logging on a sustainable basis. There are now 59 reserves in 29 tropical countries alone, and several of these biosphere reserves are important centers of research on tropical ecosystems.

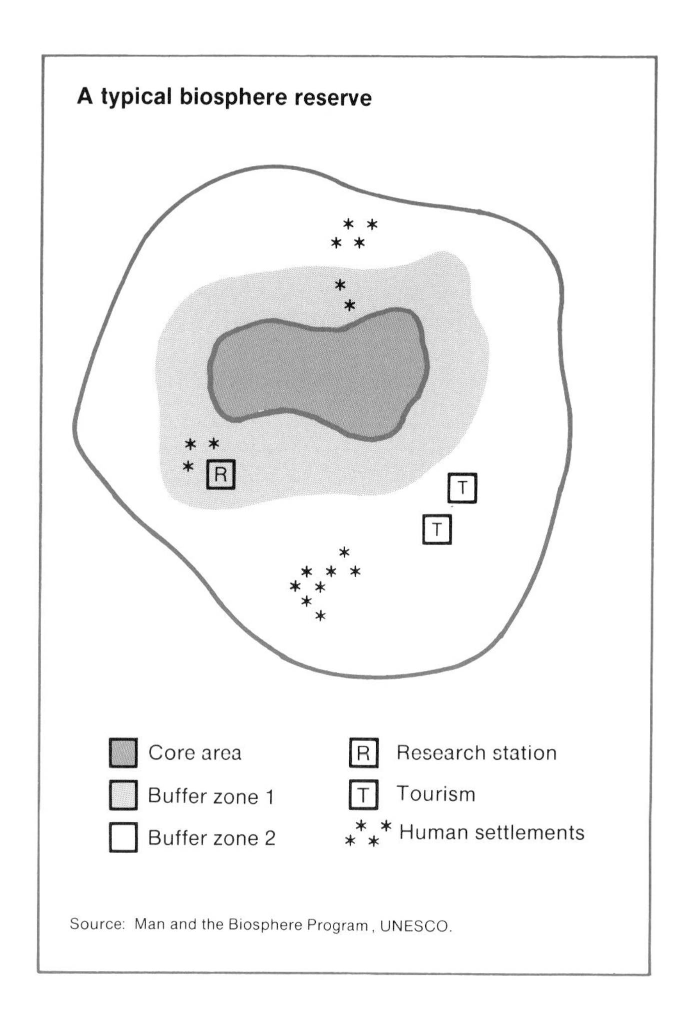

International conventions

• World Heritage Sites are designated by convention. They protect natural features considered to be of outstanding and universal value. So far, 57 "natural" sites have been designated under the convention, and 25 are in tropical countries.

• Important wetland areas have been recognized: 301 sites have been designated by 37 countries under the Convention on Wetlands of International Importance Especially as Waterfowl Habitat. Some of these sites are in or adjacent to tropical forests and protect critical ecosystems such as mangroves.

• The Convention on International Trade in Endangered Species of Wild Fauna and Flora (CITES) has monitored threatened and endangered species since 1975. The convention restricts trade in many tropical forest species, and 88 nations are now party to the convention.

Conservation data centers and national conservation strategies

• Conservation data centers collect and analyze information on critical ecosystems and species of a country or region. The data are used in setting conservation priorities and guiding land-use planning. Global databases on species and protected areas are maintained by the IUCN Conservation Monitoring Centre. National conservation data centers now exist in five tropical countries, and 24 others are planned.

• More than 40 National Conservation Strategies exist or are being developed worldwide. They are an outgrowth of the World Conservation Strategy developed by IUCN, the World Wildlife Fund, and the United Nations Environment Programme.

Protected areas to conserve tropical ecosystems

• The endangered mountain gorilla and its habitat are being protected in Parc National des Volcans in Rwanda, one of the world's poorest and most densely populated countries. Besides protecting a critically threatened species, the park protects an important watershed. Water yields are higher from the region because of the forest areas protected within the park, and farmers can reap multiple harvests including during the summer dry season. The development of the national park has received strong support from a consortium of international and developed country nongovernmental organizations.

• In Panama, 30,000 Kuna Indians live in more than 60 villages on the small, near-shore islands of the Caribbean coast. Working with the government and development assistance agencies, they have developed boundaries, multiple-use areas, and management plans for a 60,000-hectare tropical forest park. The site includes a core area of undisturbed, primary forest. Facilities are being built to carry out scientific research and to encourage tourism in the area. The Kuna are creating an environmental education program for their own people and they are helping researchers with studies of traditional agroforestry techniques.

• The Dumoga-Bone National Park, on the island of Sulawesi in Indonesia, covers 300,000 hectares. Over 90% of the park is primary tropical rain forest. It contains populations of Sulawesi's protected endemic mammals and many of the island's 80 endemic bird species. A major international research project is now underway. The Dumoga Irrigation Project was designed originally to increase rice production in the area; the park was planned with the irrigation project to protect the watershed. The World Bank channeled funds to international nongovernmental organizations (IUCN and WWF) to support the conservation elements of the project—an excellent example of the way nongovernmental organizations can complement the programs of development assistance agencies.

• Nonhuman primates are important in biomedical research. By 1970, there were signs of a shortage in animal stocks for research. The Pan American Health Organization, with the governments of Peru, Brazil, and Colombia, began maintaining a captive primate breeding center near Iquitos, Peru, taking genetic stocks from tropical forests in the region. The captive breeding program has helped curtail illegal trade in the primate species. As an outgrowth of the project, Peru is now collaborating with several groups to develop management plans for the Pacaya-Samiria National Reserve, and two additional sites are being evaluated to ensure protection of primates and other tropical forest wildlife.

Proposals for accelerated action

General strategy and policy issues

Tropical forests can only be conserved if the human pressures on them are reduced. Economic development and expanded social services are needed to improve the quality of life of the people who live in and around threatened forest reserves.

Natural forest management can yield many nontimber products of value to local populations and should be practiced wherever possible, especially to safeguard the way of life of indigenous peoples.

The full range of conservation techniques and designations should be used along with strictly protected zones or core areas. Multiple-use zoning and other land management techniques are more appropriate for the great majority of conservation units within tropical forests. Privately-owned lands and lands under lease agreements can contribute to the global system of protected areas. Incentives such as property tax relief and management agreements should be used to encourage private sector involvement. Good examples exist in India, Great Britain, Australia, Malaysia, South Africa, and the United States.

Governments should accede to the major international conservation conventions. They should endorse and implement the World Conservation Strategy, the Bali Action Plan, and the Biosphere Reserves Action Plan. Tropical forest conservation is a major goal of these international instruments.

Nongovernmental organizations are often more effective than government agencies in establishing and managing conservation areas. Their role in managing and conserving tropical forests should be recognized, expanded, and supported by development assistance agencies and national governments.

Off-site *(ex-situ)* conservation techniques, such as zoos, botanical gardens, and gene banks, are essential to conservation. Their use should be expanded. However, conservation areas serve as sources of the biological raw materials for all these activities, and therefore the highest priority must continue to be placed on protecting intact tropical forest ecosystems.

Governments and development assistance agencies should consider withholding investments from highway projects, hydroelectric development, and settlement programs in undisturbed tropical forest. Large resettlement schemes and related economic incentives often work against forest conservation and development. They should be used only when it is clear that degradation of primary forest will not result.

To apply this strategy, these important policy issues need to be addressed:

- There must be greater government commitment to conserving tropical forest resources. This commitment will come as conservation awareness increases and as the real costs of forest loss are made clear.
- Development agencies must consider conservation priorities in project planning, recognizing that conservation and development can be complementary.
- Government agencies need to be strengthened considerably. For example, more trained personnel are needed in most developing country departments of forestry, parks, and wildlife.
- Attention must be paid to the intangible benefits of tropical rain forests. The remoteness of these forests from the daily life of most people leads to public skepticism about the value of investment in their conservation. Difficulties in quantifying the long-term economic costs of tropical rain forest destruction will be the most intractable issue to resolve.
- Unregulated migration and national resettlement schemes continue to cause widespread forest loss. Governments must try to channel population growth and movement into areas already deforested.

Recommended actions for a five-year program, 1987–91

- Reduce pressures on tropical forests by intensifying agriculture on nonforest lands; incorporating trees into farming and pastoral systems; and establishing plantations on degraded, already cleared land rather than cutting undisturbed forest.
- Expand substantially the global system of conservation areas in tropical countries. All remaining tropical moist forests should be managed to ensure their long-term conservation and incorporation into national land-use plans. Only a few of the new conservation units designated should be national parks. Other types of zoning and management—such as forest reserves and wildlife management areas—should be used more widely.
- Establish additional conservation areas to include centers of origin of wild crops and other useful plant species; centers of high biological diversity; centers of high species endemism; areas considered suitable for long-term ecological monitoring and related ecosystem research in the tropics; and areas of primary tropical forest under extreme threat of conversion or destruction.
- Take immediate steps to minimize or eliminate further destruction or conversion of national parks and other conservation areas identified as being under severe threat of unlawful exploitation or encroachment. Examples of high-priority areas are given in the table.

Africa	*High-priority areas*
Cameroon	Korup National Park (proposed) Dja National Park (proposed) Pangar-Djerem National Park (proposed) Mt. Cameroon
Gabon	Forests in Gabon-Cameroon border region All protected areas
Ivory Coast	Tai National Park Mt. Nimba Nature Reserve (region)
Liberia	Sapo National Park Mt. Nimba Nature Reserve (region)
Madagascar	All forest on east side of island
Zaire	Garamba National Park Forests bordering Rwanda
Asia	
India	Andaman and Nicobar Islands Gir National Park Silent Valley National Park Manas Tiger Reserve Mangrove ecosystems Western Ghats (several areas)
Indonesia	Irian Jaya (several areas) Kutai Game Reserve Kalimantan (several areas) Siberut Kerinci-Seblat National Park
Papua New Guinea	Western PNG—Fly to Sepik Rivers Huon Peninsula
Philippines	Mt. Apo National Park
Thailand	Lowland forests on Malay Peninsula Thung Yai Wildlife Sanctuary Huai Kha Khaeng Wildlife Sanctuary
Latin America	
Bolivia	Entire protected areas system
Brazil	Southeast Atlantic Coastal forests Araguaia National Park Amazon basin: several areas

Colombia	Choco region Sierra Nevada de Santa Marta Amazonia
Costa Rica	La Amistad International Park Zona Protectora La Selva Corcovado National Park
Ecuador	Cuyabeno and Curaray Reserves Pajan and Paute Protection Forests Mangrove ecosystems Yasuni Research Station Amazonia Coastal forests of northwest
Nicaragua	Saslaya National Park Mosquitia forests near Honduras border
Panama	Darien National Park La Amistad International Park Kuna Indian Forest Park
Peru	Manu National Park Amazonia Coastal forests (Loma formation)
Venezuela	Peninsula de Paria National Park Sierra de Imataca Altiplanicie de Nuria Laguna de Tacarigua National Park

- Develop a national conservation strategy in most developing countries.
- Develop conservation data centers building on models that exist in Peru, Costa Rica, Puerto Rico, and Colombia. Formal links should be developed between the data centers, government agencies, international development assistance agencies, and nongovernmental organizations, especially IUCN's Conservation Monitoring Centre.

Investment needs

The estimated cost of a 5-year investment program for the 21 countries listed in the table is US$548 million. Investments in the following categories are proposed:

Development of sustainable agriculture near existing conservation areas to relieve pressure on tropical forests

Improvement of protection and management of existing conservation areas

New conservation units in various categories

National conservation strategies

National conservation data centers

Building national parks and wildlife agency capability

Conservation education, training, and extension

Conservation of tropical forest ecosystems

Summary of needed investments, 1987–91

Africa	*Million US$*
Cameroon	30.5
Gabon	12.7
Ivory Coast	23.7
Liberia	13.0
Madagascar	24.9
Zaire	23.8
Asia	
India	32.2
Indonesia	42.7
Malaysia	34.3
Papua New Guinea	10.0
Philippines	30.4
Thailand	27.7
Latin America	
Bolivia	31.3
Brazil	50.0
Colombia	30.0
Costa Rica	20.5
Ecuador	17.1
Nicaragua	17.4
Panama	20.9
Peru	35.6
Venezuela	19.6
Total (21 countries)	548.3

Strengthening institutions for research, training, and extension

The problem

In many developing countries, two of the most formidable constraints to expanding forestry are weak research programs and shortages of trained forestry personnel, including extension workers.

Because of inadequate data, weak monitoring capabilities, insufficient operating funds, and the shortage of trained personnel, many forestry administrations are unable to implement policies and effectively plan and manage research, training, extension, and other forestry programs.

Recruiting and retaining highly qualified and motivated researchers, teachers, field managers, and extension agents is often difficult, because of poor career opportunities, low prestige attached to forestry, and low salaries. In all regions, there are shortages of trained forestry staff, particularly at the vocational and technician levels. FAO has estimated that Latin America has the capacity to train a sufficient number of professional foresters, but in 1980 the region had a shortage of 12,000 forestry technicians.

Forestry training institutions have been neglected in many countries. Curricula for education and training must be revised to reflect the changing needs and priorities in forestry, particularly the emerging emphasis on farm and community forestry and local participation. Foresters, traditionally trained to protect government forest reserves and to manage them for industrial wood production, generally lack experience in working with local people and community groups and often are insensitive to their needs. There is an urgent need to increase the amount of practical training, to improve teacher training and training techniques, and to modernize and expand training centers.

Most forestry research institutions, particularly in Africa, are weak. They have shortages of trained researchers, equipment, and operating funds. As with training, forestry research priorities must be made more relevant to the problems facing developing countries.

Technology transfer is poor because of weak links among research, training, and extension and the lack of information sharing among countries. Few countries have well developed forestry extension services, and too often foresters are drawn to urban areas where career opportunities are greatest.

Success stories

Research

In temperate regions, forestry research has resulted in higher yield and commercial value of tree species through genetic improvement of planting stock and better management practices. Similar benefits could be expected for many tropical tree species on which research to date has been limited.

- The potential for higher yields from plantations through genetic improvement of trees has been shown dramatically by Aracruz Florestal, a Brazilian paper company. The first step in raising the yield of its eucalyptus plantations was to import seed from Africa to replace low-yielding trees in Brazil. Researchers then

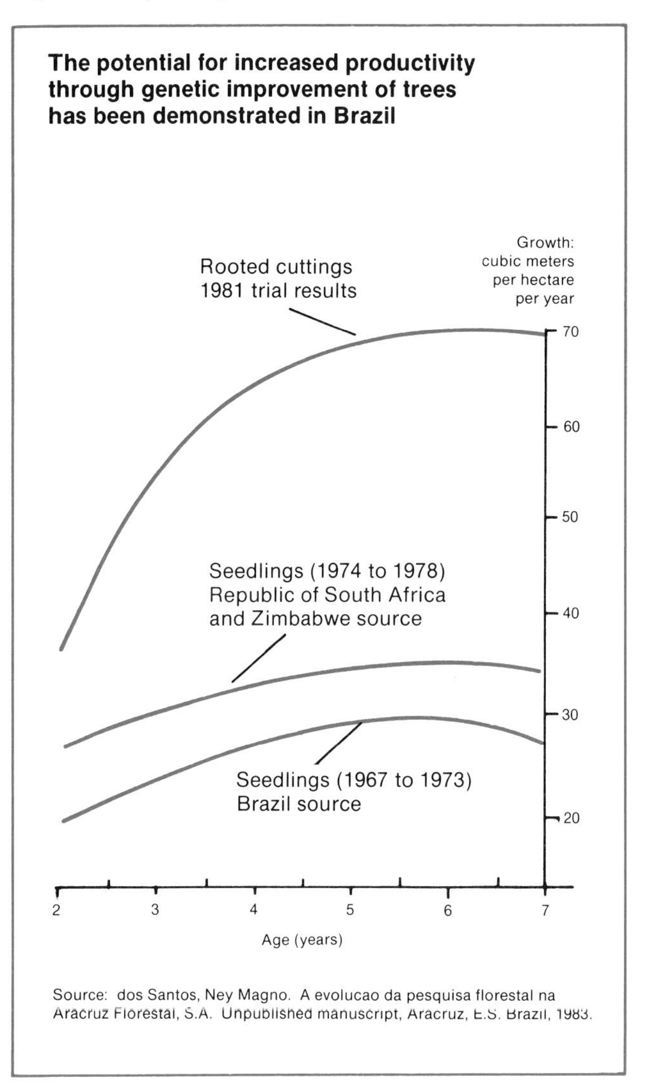

The potential for increased productivity through genetic improvement of trees has been demonstrated in Brazil

Source: dos Santos, Ney Magno. A evolucao da pesquisa florestal na Aracruz Florestal, S.A. Unpublished manuscript, Aracruz, E.S. Brazil, 1983.

selected genetically superior trees, cloned them, and produced better planting stock, doubling plantation yields from 33 to 70 cubic meters per hectare per year.

• Carton de Colombia, a forest products company in Colombia, pioneered research in the 1950s and 1960s on technologies for pulping mixed tropical hardwoods. At the time, Colombia faced a rising import bill for paper products, yet its large hardwood resource scarcely was being tapped commercially because the great diversity of tropical hardwood species made their use difficult. Pulping technologies were available for temperate softwood species but not for tropical hardwoods. The company's research on processing technologies has enabled it to use more than 265 tree species for pulping and to develop a large and successful pulp and paper industry important to Colombia's economy.

• By investing in wood utilization research over the past 20 years, Malaysia has been able to increase the number of tropical forest tree species it uses commercially from 100 to more than 600.

Research networks among countries can be highly useful for coordinating research, sharing information and materials, and carrying out training. Although well developed for agriculture (such as the networks of the Consultative Group for International Agricultural Research), only a few narrowly focused networks exist for forestry (for lowland tropical pines, eucalyptus, poplars, leucaena).

• The Commonwealth Forestry Institute (CFI) in Oxford has promoted the worldwide distribution and use of Central American tropical pines through its research network. When CFI started its work in 1962, large plantations of these species were being established with seed collected from only a small portion of their natural ranges, thus failing to take advantage of the species'

Research networks have helped bring about worldwide distribution and use of Central American tropical pines

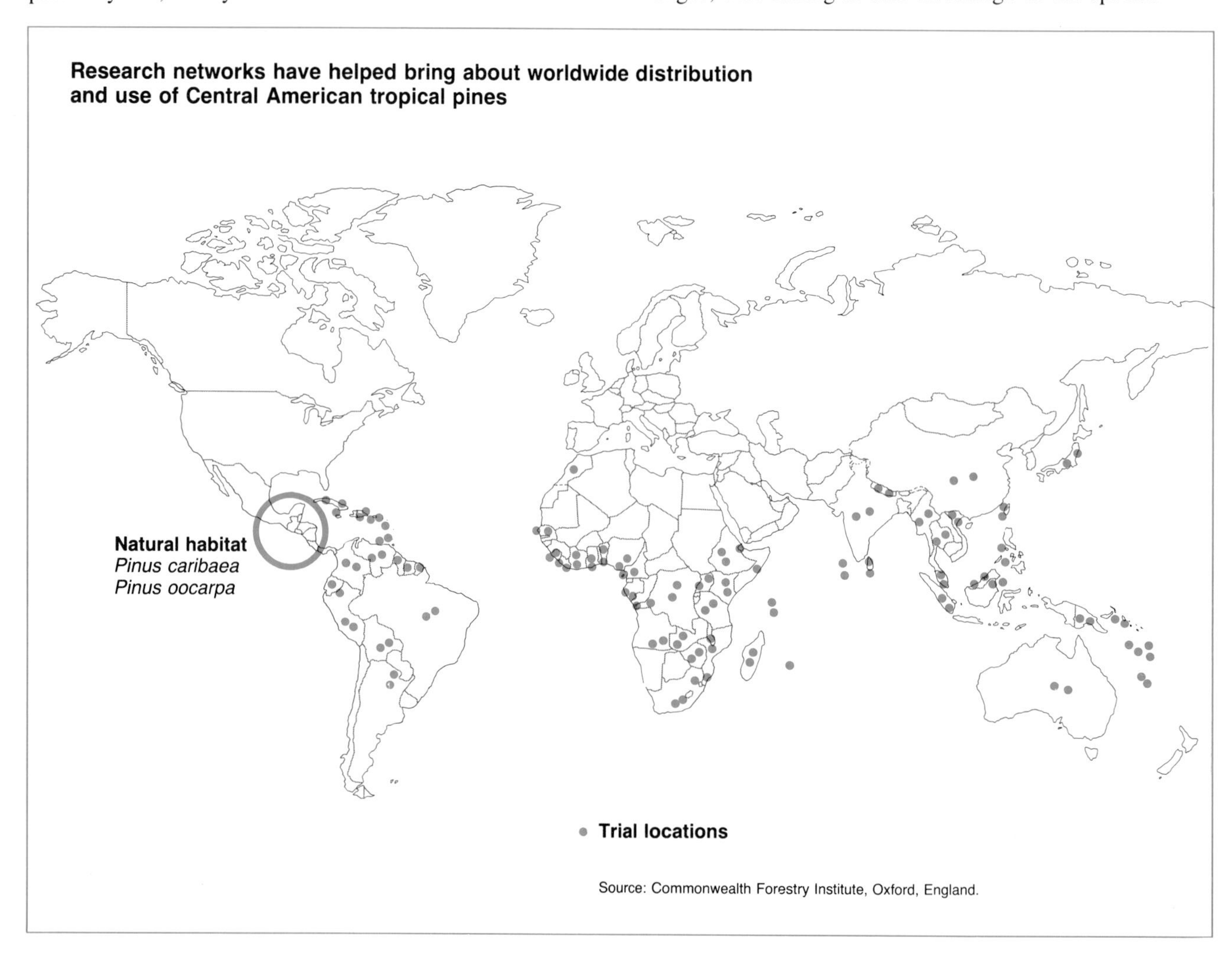

Source: Commonwealth Forestry Institute, Oxford, England.

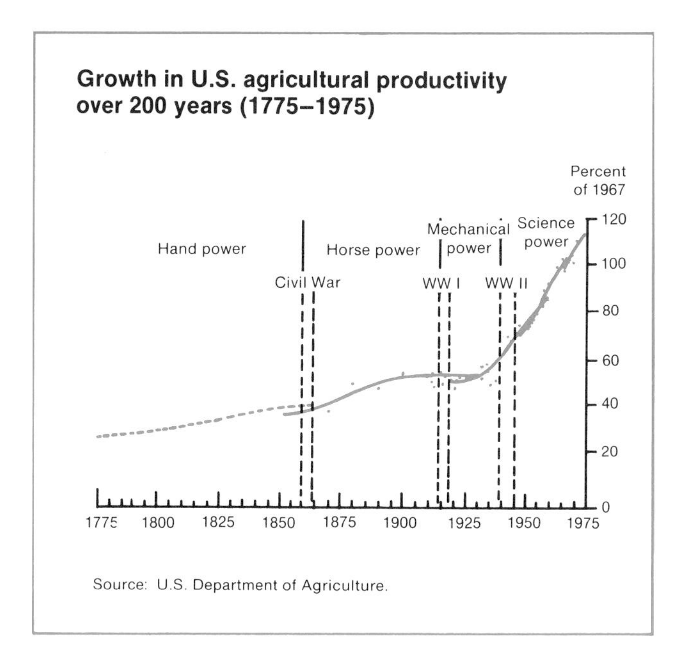

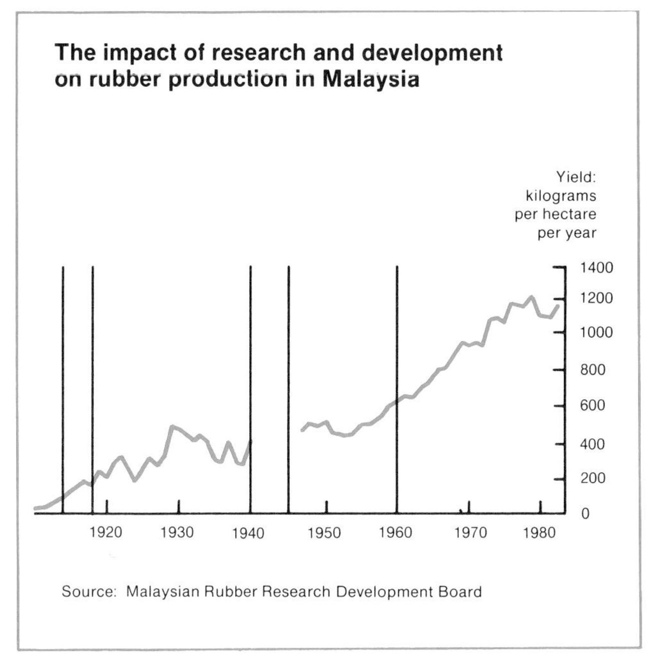

genetic variability. Through the networks, CFI has helped collect, distribute, test, and evaluate various genetic stocks of these pines. Hundreds of trials of two species have been carried out in 50 tropical countries. By exchanging seed, standardizing experimental design and assessment methods, and developing data management systems for information sharing, the networks have enabled countries to match the genetic material to a site, thereby increasing plantation yields.

Research has had a key role in agricultural development in temperate and, more recently, in tropical countries. Forestry can learn important lessons from agriculture.

- The most rapid increases in U.S. agricultural yields have occurred since 1930 as a direct result of intensified research and development.
- A 5.5-fold increase in rubber yields in Malaysia between 1920 and 1980 resulted from planting improved stock and better management, both made possible by research.
- International agricultural research centers and networks have had a great impact on agricultural yields in developing countries. Concentrated research efforts brought about the "green revolution," best shown by dramatic increases in wheat and rice yields.

Training

Many forestry training institutions have been established in the past few decades. Today, 190 institutions in Asia, 104 in Latin America, and 87 in Africa provide forestry training at the university or technical level. Most are new and small and still face serious problems, but some effective national and regional institutions have been developed.

- In the past 22 years the Department of Forest Resource Management at the University of Ibadan, Nigeria, has grown from 3 small rooms to a well equipped facility that is one of West Africa's leading institutions offering postgraduate forestry training. By 1979, it had graduated 223 B.Sc., 15 M.Sc., and 13 Ph.D. students, thus greatly strengthening Nigeria's capabilities in forestry.
- The College of African Wildlife Management in Mweka, Tanzania, serves anglophone Africa and has trained more than 1000 wildlife managers since 1963. Virtually all protected areas in east Africa employ Mweka graduates, and some graduates have attained high-ranking government positions. Mweka has served as a model for newly established national and regional wildlife schools in Africa and Asia.
- The National School for Forestry Sciences in Honduras is the only institution in Central America that provides training for forest technicians. Since 1969, it has graduated more than 500 students, conducted numerous short courses for regional forestry personnel, built an extensive physical plant, earned a reputation for high academic standards, and created a world-renowned seed bank.

Students at this forestry training center in Ecuador will go on to work for the Forestry Service.

A forestry extension agent demonstrates a chainsaw to members of a tree-growing commune in China.

• The Tropical Agricultural Research and Training Center in Costa Rica serves Central America and the Caribbean. It has trained more than 275 master's level students and provided specialized short-term training to more than 1000 professionals from the region. Its research programs have helped to improve research capabilities of national institutions and to strengthen regional cooperation. In-service training courses provide essential updating of skills and short-term training in areas not yet well developed in technical schools and universities. The center recently has given national and regional short courses in watershed management, agroforestry, park management, fire prevention, and fire fighting.

• The East-West Center's Environment and Policy Institute in Hawaii has extensive training and research programs in 5 program areas, including forest lands policy and water resources management. Over the past 10 years, the institute has hosted and trained hundreds of participants from Southeast Asia.

• The University of Michigan's International Seminar on National Parks and Other Protected Areas, sponsored by the U.S. National Park Service and Parks Canada and now in its 19th year, is a 4-week mobile course that provides senior-level park administrators from around the world an opportunity to compare various forms of park administration and management in North America. Its success has led to a similar course for forest managers from the developing world.

Twinning a developing country institution with one in another country is a cost-effective way to train staff and students, transfer information, and build up management capabilities. Examples in forestry include the State University of New York and the University of the Philippines; the University of Freiburg, Germany, and Curitiba, Brazil; and the University of Toronto and Universidad Nacional Agraria, Lima, Peru.

Extension

While effective forestry extension services in developing countries are still rare, several countries—such as Nepal, Kenya, India, South Korea, and Burkina Faso—recently have developed forestry extension programs.

• Extension has been an essential part of Nepal's Community Forestry Programme. A large number of field staff working locally and using written and audio-visual materials and mass media such as radio, have effectively promoted community forestry efforts and provided technical support to participating villages. Between 1980 and 1984, extension was important in establishing and managing 400 nurseries, planting 7500 hectares of plantations, and installing 6000 stoves.

• India has made considerable progress in adding forestry to agricultural extension training. Forestry staff have organized short courses in forestry techniques for village level agricultural staff who provide forestry advice to farmers and communities.

Proposals for accelerated action

General strategy and policy issues

Concentration of effort is recommended in 6 main areas. National governments and development assistance agencies should—

Strengthen the capabilities for policy formulation and planning by national forestry administrations.

Improve integration of agriculture and forestry in research, training, and extension through collaborative research programs, revised curricula, and restructuring of extension programs.

Emphasize agroforestry, socioeconomic factors in forestry and land use, low-cost technologies for rural forestry programs, and extension techniques in training programs.

Strengthen national research, training, and education institutions and develop stronger links between national and regional institutions.

Concentrate on a few high-priority research topics with high potential impact on rural poverty.

Involve local people in extension and outreach programs.

To apply this strategy, these important policy issues need to be addressed:

- Steps must be taken to raise the generally low political and financial commitment of national governments to forestry institutions.
- High-quality, committed staff must be recruited and retained; to do so will require that national governments upgrade career paths and reward structures for all levels of personnel.
- Governments need to examine incentive programs to encourage personnel to seek posts in rural rather than urban areas.
- A general skepticism toward research will have to be reversed before research will receive adequate financial support.
- The building of strong extension programs requires a commitment to retraining personnel on all levels; making structural changes in forestry, agriculture, or rural development departments; and allocating adequate financial resources to extension activities.
- Regional training and research institutions fulfill essential roles, but they typically suffer from serious financial constraints. Sustained international funding is essential to the continued development of such institutions until long-term financing can be found. Endowments should be used more widely to gain some measure of financial security.

Recommended actions for a five-year program, 1987–91

National activities

Sector planning and project preparation

- National governments, with the help of development assistance agencies, should carry out sector planning studies.
- The sector studies should examine the organizational structure and capacity of the forest department; forestry and related resource policies; research and training needs; and legislation that affects forest resources. Forestry codes, land and tree tenure laws, as well as agricultural, land use, and rural development laws should be revised to include incentives for tree growing at the farm and community level.
- Following these sector reviews, national governments should be helped to prepare and appraise forestry projects that could be suitable for support by external aid agencies. FAO's Investment Centre has played a leading role in this area in the past and it should be further strengthened to carry out this important work.

Research priorities

- Fast-growing, multipurpose tree species, including selection and breeding research to maximize sustainable yield of the desired products
- Biophysical and socioeconomic research on incorporating trees into farming and grazing systems
- Improved use of wood and nonwood products from native and introduced species including use of lesser known species, logging wastes, and minor forest products, ranging from appropriate small-scale technologies to large-scale, capital intensive applications
- Natural forest management for wood and nonwood products, including silvicultural research, harvesting research, multipurpose management of savanna woodlands, and basic ecological research
- Forest inventory, monitoring, and resource analysis.

Training

- Strengthen national training capabilities to meet needs for trained personnel at all levels.
- Strengthen regional training institutions to serve smaller countries without national facilities or to provide training in special subjects.
- Organize short courses to change attitudes and upgrade skills.
- Improve the content of forestry, agriculture, and natural resources training programs to put greater emphasis on agroforestry, fuelwood production systems, integrated watershed management, natural forest management, and conservation.
- Provide more fieldwork in training programs.
- Support twinning relationships between universities in developing and developed countries and similar links between universities in developing countries.

Extension

- Develop forestry extension capability within the forestry, agriculture, or rural development departments and ensure that it is adequately linked with research and training institutions.
- Develop extension materials, such as pamphlets, flipcharts, and audiovisual presentations.
- Increase the use of mass media and other outreach mechanisms to raise public awareness of the importance of forestry, to promote forestry as a rural development activity, and to provide useful technical information to local people.
- Increase participation of private voluntary organizations, community groups, and schools in extension work.

Regional and international activities

- Strengthen professional and technical training programs for Africa, including the new program in professional forestry training for warm regions at the National School of Waterways and Forest Rural Engineering in Montpelier, France, and the proposed Natural Resources Institute in the Ivory Coast.
- Expand agroforestry and farming systems research in Africa, particularly the programs of the International Council for Research in Agroforestry in Nairobi.
- Increase financial support for regional training institutions to cover operating costs and to provide scholarships.
- Fund the 5-year regional training project in Latin America proposed by the World Wildlife Fund. The project would review training requirements in natural resources and environment, guide technical and financial cooperation, support individual institutions, and sponsor activities to meet the needs of the region.
- Establish a separate fund within development assistance agencies to which developed and developing country universities and research institutions could apply for funds for twinning relationships, scholarships, conferences, and workshops.
- Strengthen FAO's ability to promote regional cooperation; to carry out regional workshops, training courses, and seminars; and to develop networks in watershed management, agroforestry, arid zone forestry, desertification control, and forest land-use assessment.
- Fund the forestry research networks being developed in Africa, Asia, and Latin America by the Special Program for Developing Countries (SPDC) of the International Union of Forestry Research Organizations. Support longer-term SPDC activities including training in forestry research management and execution, information transfer to developing countries, twinning arrangements, and an international fund for forestry research and training.
- Provide adequate financial support for the UNESCO Man and the Biosphere (MAB) Programme, particularly MAB's Project 1 on Ecological Effects of Increasing Human Activities on Tropical and Sub-Tropical Forest Ecosystems and Project 8 on Biosphere Reserves.
- Strengthen FAO's ability to develop extension materials, to advise regional instructors on designing extension programs, and to develop teaching materials and provide training workshops on forestry extension (such as the UNDP/FAO Asia and Pacific Programme for Development Training and Communication Planning in Bangkok).
- Support efforts to increase the participation and effectiveness of developing country nongovernmental organizations involved in tree planting and related rural development work (such as The Tree Project, based in the U.N. Non-Governmental Liaison Office, and the Environment Liaison Centre in Nairobi).

Summary of needed investments, 1987–91

International development assistance agencies generally allocate between 15 and 25% of their total forestry investment to research, training, and extension. Twenty percent of the total investments recommended here, or $1,064 million over the next 5 years, should be invested in institution strengthening activities. About 15-16% should be allocated to strengthening national institutions, while the remaining 4-5% should be devoted to the regional and international activities outlined above.

Strengthening institutions for research, training, and education

Summary of needed investments, 1987–91

Million US$

Activity	*Africa*	*Latin America*	*Asia*	*Total*
Fuelwood and agroforestry	68	100	155	323
Land use on upland watersheds	52	30	233	315
Forest management for industrial uses	43	141	140	324
Conservation of forest ecosystems	25	48	29	102
Total	188	319	557	1,064

Summary of total investment needs

Estimated costs* in million US$ for the period 1987–91

	Fuelwood & agro-forestry	Land use on upland watersheds	Industrial forestry	Ecosystem conservation	5-year totals
Africa					
Botswana	15	—	—	—	15
Burkina Faso	25	—	—	—	25
Burundi	20	—	—	—	20
Cameroon	—	—	20	31	51
Cape Verde	15	—	—	—	15
Chad	14	—	—	—	14
Congo	—	—	20	—	20
Ethiopia	40	100	—	—	140
Gabon	—	—	—	13	13
Ghana	—	—	10	—	10
Ivory Coast	—	—	75	24	99
Kenya	48	35	—	—	83
Lesotho	10	—	—	—	10
Liberia	—	—	15	13	28
Madagascar	30	10	—	25	65
Malawi	24	—	—	—	24
Mali	30	—	—	—	30
Mauritania	16	—	—	—	16
Niger	20	—	—	—	20
Nigeria	50	—	35	—	85
Rwanda	30	—	—	—	30
Senegal	25	—	—	—	25
Somalia	15	—	—	—	15
Sudan	35	—	—	—	35
Tanzania	30	—	—	—	30
Uganda	15	—	25	—	40
Zaire	—	—	10	24	34
Zimbabwe	—	46	—	—	46
Subtotal	507	191	210	130	1038
Asia					
Bangladesh	52	—	—	—	52
Burma	—	—	30	—	30
China	250	135	285	—	670
India	500	500	190	32	1222
Indonesia	—	100	50	43	193
Malaysia	—	—	40	34	74
Nepal	30	15	—	—	45
Pakistan	40	45	20	—	105
Papua New Guinea	—	—	15	10	25
Philippines	—	120	40	30	190
Sri Lanka	30	—	—	—	30
Thailand	—	—	35	28	63
Subtotal	902	915	705	177	2699
Latin America					
Argentina	—	—	100	—	100
Bolivia	25	—	—	31	56
Brazil	400	10**	325	50	785
Chile	—	—	50	—	50
Colombia	—	50	45	30	125
Costa Rica	15	—	15	21	51
Ecuador	—	15**	20	17	52
El Salvador	10	—	—	—	10
Guatemala	—	—	15	—	15
Haiti	15	—	—	—	15
Jamaica	—	10**	10	—	20
Mexico	—	—	90	—	90
Nicaragua	—	—	—	17	17
Panama	—	20**	—	21	41
Peru	25	20**	30	36	111
Venezuela	—	—	25	20	45
Subtotal	490	125	725	243	1585
Total	1899	1231	1640	550	5320***
	36%	23%	31%	10%	100%

*All totals rounded
**Preliminary estimate, pending additional research
***Approximately 20% of this investment would be allocated to research, training, education, and extension.

Photograph credits

All photographs in this report are from the archives of the United Nations Food and Agriculture Organization except the National Geographic Society's photograph on page 12 and the photograph of the Henri Pittier National Park in Venezuela on page 34 by F. William Burley.